Specifying Architecture

a guide to professional practice

John Gelder

ISBN 0 9586187 4 7

First edition 1995

Second edition 2001

Printed by Focus Press

Published by Construction Information Systems Australia Pty Ltd
61 Lavender Street, Milsons Point, New South Wales 2061
Telephone (02) 9923 1499, fax (02) 9955 3537
Internet mail@cis.asn.au & http://www.cis.asn.au

Construction Information Systems is a not-for-profit company founded in 1975 to improve communications in the building industry. Its shareholders are 20 leading stakeholders in the industry:

- *Air-Conditioning and Mechanical Contractors Association of Australia*
- *Association of Consulting Engineers Australia*
- *Australian Council of Building Design Professions*
- *Australian Elevator Association*
- *Australian Institute of Building*
- *Australian Institute of Quantity Surveyors*
- *Building Control Commission, Victoria*
- *Construction Industry Engineering Services Group*
- *Department of Administration and Information Services, South Australia*
- *Department of Finance and Administration, Australia*
- *Department of Housing and Works, Western Australia*
- *Department of Public Works, Queensland*
- *Department of Public Works and Services, New South Wales*
- *Department of Transport and Works, Northern Territory*
- *Department of Treasury and Finance, Tasmania*
- *Department of Urban Services, Australian Capital Territory*
- *Institution of Engineers, Australia*
- *Master Builders Australia*
- *Royal Australian Institute of Architects*
- *Standards Australia*

Cover photo of the National Museum of Australia: John Gollings
Architects: Ashton Raggatt McDougall/Robert Peck von Hartel Trethowan

Contents

Preface

Specifications are the essence of things hoped for, the image of things not seen. A perfect specification writer should have graduated from Columbia Law School and Drummond's Detective Agency and then taken a course in palmistry to cover the unforeseen contingencies (Thumtack 1914).

This book is based on the supposition that building specifications are worthwhile. Decent specifications can save time, heartache, confusion, frustration, money and (occasionally) reputations. That buildings get built without the benefit sometimes of a decent specification is a tribute to the willingness and tenacity of those involved in the construction process. With a little extra effort by specifiers, their work – and that of many others in the industry – could be made a lot easier.

However, it seems that the industry has found it difficult to prepare adequate specifications, and other documents, for over 150 years:

Almost solely from sufficient pains not being taken in drawing the specifications for buildings, and from a want of proper foresight, may be traced, most of the disputes between the builder and the architect and the employer, which so often occur, and which lead to lawsuits and arbitrations, which are ofttimes so excessively and even ruinously expensive, and though final are unsatisfactory to all parties (Bartholomew 1846).

At least half the troubles in the architectural world are due to faulty and ill-expressed documents (Eggleston 1955).

The Research Report [of November 1988] argued that errors, contradictions, ambiguity in and late delivery of contract documentation, giving rise to delays and inefficiencies, are the greatest cause of claims and disputes in the construction industry ... The Report also refers to the late issue of documents particularly in relation to further documentation provided during the construction period. A view has been expressed that the quality of documentation has been reducing over recent years with a lowering of the level and detail of documentation standards (NPWC/NBCC 1990).

In February 1994, what is now the Department of Public Works and Services (DPWS), NSW,[1] advised specifiers, both internal and external, that technical specifications "often contain inappropriate and redundant requirements". Further, they introduced unwanted risks on the part of the principal, were often out of date, and defined roles for the contract administrator that conflicted with current policies.

The problem is not confined to Australia. In the UK, the Construction Industry Board's working group 11 (concept subgroup), in response to Sir Michael Latham's *Constructing the Team* report, identified the main areas of avoidable cost as "changes, overspecification, inappropriate specification, waste and duplication, unnecessary complexity, conflict and programming" (Thompson 1996).

Finally, a survey by CSIRO found that deficiencies in documentation have been steadily increasing over the past 12-15 years, leading to poorer quality projects and an increase in overall project costs. Consultants blame this state of affairs on the reduction in design fees (Tilley and McFallen 1999).

This book is an attempt to set out best-practice professional guidelines for the preparation and use of contemporary specifications, and to go some way to remedying the above longstanding concerns. Rehashed versions of books on the subject originally written in the 1950s are of little help here (Whiteley 1998). The context and practice of specifying has changed completely over the past 50 years. Before going too far we need to define what is meant by a specification. Here are a couple of definitions that have been around for a while:

Specification: A detailed description prepared by a consulting engineer or architect to tell the contractor everything about the workmanship which cannot be shown on the drawings. It is written in the sequence of trades (Scott 1964).[2]

Specification: The document that prescribes the requirements with which the product or service has to conform.

NOTE – A specification should refer to or include drawings, patterns or other relevant documents and should also indicate the means and the criteria whereby conformity can be checked (ISO 8402:1986 Quality vocabulary).[3]

These conventional definitions of the specification suggest that for specifiers the primary function of the specification, and indeed the drawings, is to define precisely what is to be built by the contractor, for the benefit of both parties to the contract. We can be more precise and suggest the following:

- Specifications, the textual or written *description* of the works, provide the definition of quality and processes.
- Drawings, the complementary graphical *description* of the works, define quantity, position, assembly and extent.

1. Formerly known as the Department of Public Works.

2. The 1984 edition has a rather more elaborate definition with a peculiarly British slant – it states that "... In some contracts with an unusually detailed bill of quantities, the specification, though provided, may be no part of the contract". The 1964 version was more useful here.

3. AS/NZS ISO 8402:1994 has a slightly different, less elegant, definition, namely a "document stating requirements".

On small projects in particular, the two are often combined in the form of extensive annotation of the drawings,[4] but this is simply not practical for larger projects. For these, the two components of the description of the works must be separated and therefore properly coordinated.[5] Almost always, *both* are necessary.

Given the blurred distinction between the drawings and specification as descriptive components of the contract, not surprisingly principles that make sense for one can often be applied to the other. For example, descriptions of items that do not exist in the project should not be provided to contractors, either as drawings or as specifications.

The specification is a pivotal document in that it links the drawings, the conditions of contract, bills of quantities and referenced documents such as standards. The specification should supplement and complement this material, not duplicate it. Information peculiar to the specification includes quality control procedures, rules for substitutions, and definitions of the required quality of manufactured products, fabricated products and built products. Quality may be defined in terms of finishes, relevant standards, material grades and thicknesses, tolerances, performance requirements, and (sometimes) methods of fabrication and installation.

The specification is also a multi-purpose document. In conjunction with the drawings, it has at least nine functions (Figure P.1), including the primary function, to define precisely what the contractor is to build.

1. Design record: statements in the specification reflect, and may be the only record of, many detailed design decisions.
2. Statutory compliance: the specification is submitted to local councils and other authorities to demonstrate proposed methods of statutory compliance.
3. Pre-tender estimates: the specification is used by quantity surveyors and others to prepare progressive cost estimates.
4. Tendering by the principal and subcontract tendering by the contractor: the specification is one of the documents tenderers use to price the work. In particular, it is disassembled for subcontract tendering.
5. Part of the contract: in this capacity the specification, with the drawings, the conditions of contract, and perhaps a bill of quantities and other documents, comprise the contract that legally binds the principal and the contractor. Indeed Standen (1995) argues that specifications (and therefore the drawings and bills as well) are nothing but a set of special conditions of contract.
6. On-site: the specification provides an on-site instruction manual and statement of requirements, and provides rules for assessing conformance of work with these requirements.
7. Disputes: arbitrators and courts will turn to the specification in a dispute, in an attempt to clarify the principal's requirements and to determine what has been agreed. In this respect, particularly if not well written, it may be a two-edged sword.

4. This annotation is still "the specification", just as a tiger in a zoo is still a tiger. Both must have teeth!

5. It has long been difficult to separate the written and graphical descriptions of the works conceptually, though. For example, the ancient Greeks used an ambiguous term, anagrapheis, which might refer to either a drawn or a written description.

8. Facility management: requirements for manuals, training and spare products e.g. luminaires, tiles will be spelt out in the specification, which also doubles as a source of information for alterations, additions and demolition, and perhaps reuse.
9. Project management: spanning these eight functions are nine project management functions – integration, scope, time, cost, quality, risk, procurement, communications and human resources. The specification addresses all of these in various ways.

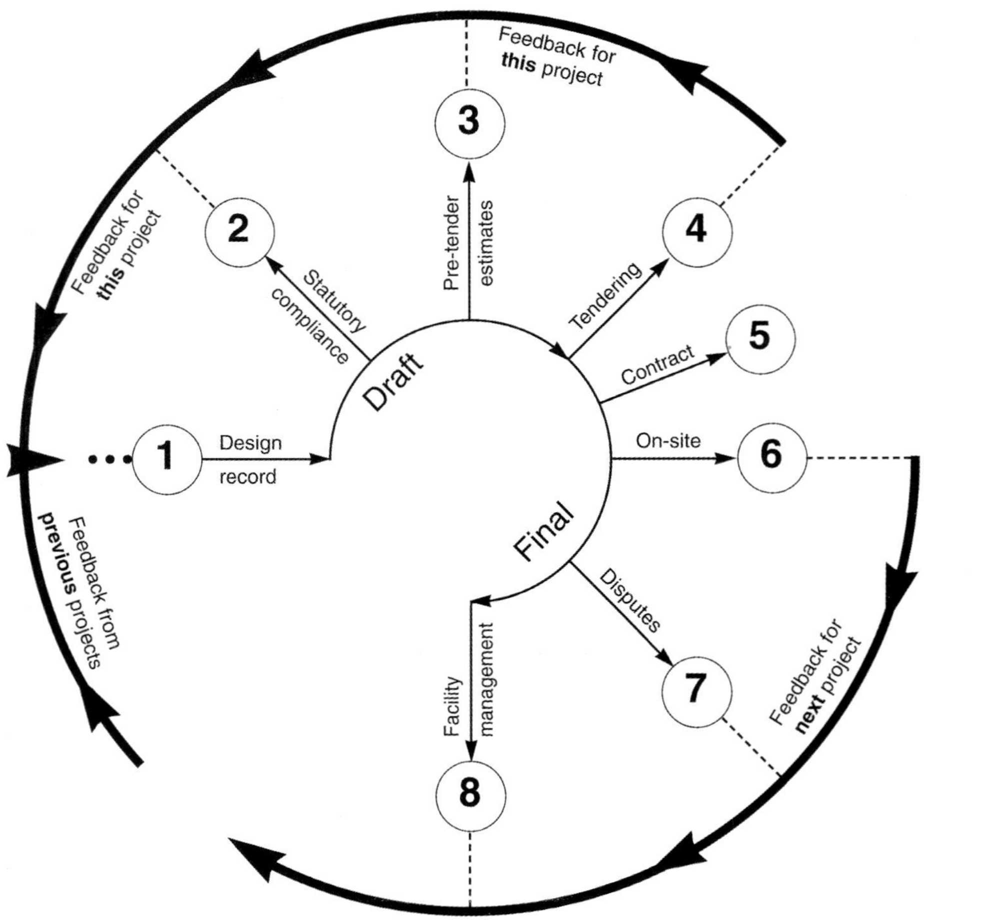

Figure P.1: Functions of the specification. The feedback loops are vital – each project must be analysed at completion and the lessons learned passed on. Everyone makes mistakes, but it is silly – and a waste of time – to repeat them. Unlike the other functions, project management is not one phase in a sequence but applies throughout so is not shown separately in the diagram.

Structure of this book

Production of a specification is a project like any other. It has a beginning and an end. It is a complex and creative task involving a multi-disciplinary team brought together for the occasion. Because it must be managed as a project, the typical project life-cycle applies to specification production:

- Inception.
- Design.
- Execution.
- Finish.

Once specifiers have finished producing the specification, others will start to use it as a product, as a tool in the execution of a separate project – the construction of the building itself. Later it will be used as a tool in the operation and maintenance of that facility (see Figure P.2).

This book examines the specification from this point of view, looking at principles and practice alongside each other. It begins with three chapters on context. Specifiers do not work in a vacuum – an understanding of context is essential. These can be skipped if you are familiar with context and if you want to get straight to the meat of the work, which deals with specification preparation (Chapters 4 to 7). Contractual matters are given extra emphasis. The book also has a chronological structure, from past to present and glimpses of the future.

But one cannot fully understand what to include in a specification, and what not to include, without understanding its use. For a complete picture, you will also need to read

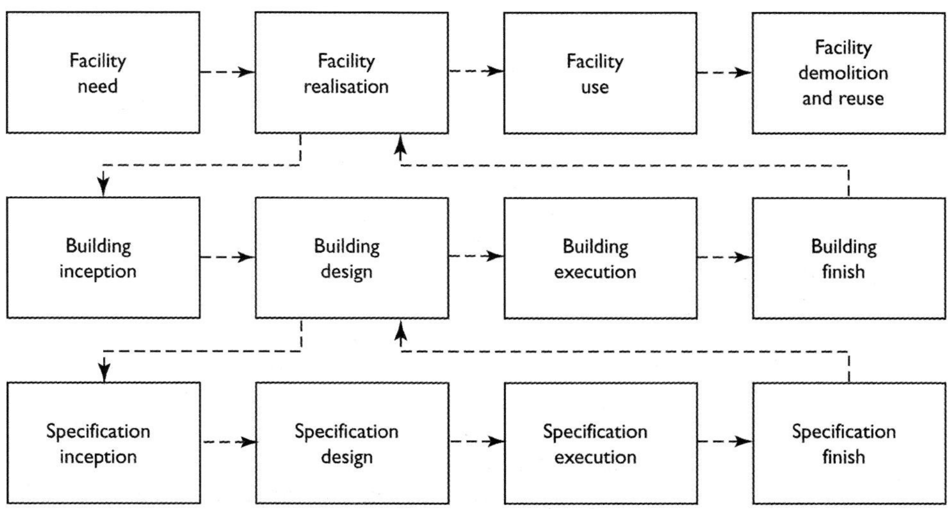

Figure P.2: Project life-cycle. The specification is a project within a project within a project!

the chapter on usage, Chapter 8. The users do not care, however, how the specification was produced. They just want a document that tells them what they need to know. It is hoped this book tells you what that is, and how to get there.

The second edition

The first edition of the book was published in 1995. Since that time a surprising amount of material has become out-of-date. Many organisations and documents referred to have changed names, disappeared or been replaced, and many new articles and books have appeared.

As well as updating this material, the second edition has provided an opportunity to expand the information in the book. A new chapter has been added on contemporary context in the United Kingdom and the United States, and the historical material has been completely revised. Contract extracts have been expanded, and extra material included on specifying construction methods, performance specifying, design and construct, use of standards, the role of manufacturers, green specifying, conformance, and plain English. It is hoped that all this helps maintain and enhance the book's relevance to Australian students and specifiers.

A note for students and teachers

The book is intended for a wide readership, from students, to part-time and full-time specifiers. Accordingly, some of the material will be too detailed for undergraduate use. It is suggested that undergraduates should be trained, at least, to prepare sound project specifications for standard residential construction, for which they should be able to

- work with the deemed-to-satisfy provisions of the domestic component of the Building Code of Australia (BCA) and domestic versions of statutory codes such as AS/NZS 3000 (2000);
- prepare tendering material for single-stage tendering;
- find and use paper, diskette and CD material from manufacturers and technical information sources;
- use a simple commercial master specification, such as *NATSPEC DOMESTIC*;
- program the creation of a simple project specification, with one specifier;
- design a simple project specification;
- coordinate with simple standard contracts, such as SBW-2;
- coordinate with drawings prepared in the specifier's office;
- coordinate worksections prepared by the specifier;
- use proprietary and descriptive methods of specifying, including management of substitution and verification;
- reference a basic set of, say, 75 Australian standards;
- prepare schedules of work;

- edit a project specification for clarity, grammar, and appearance;
- produce a specification on paper and disk;
- use a specification for design record, tendering, contractual, and on-site purposes;
- specify for construct-only procurement; and
- prepare sound traditional domestic worksections.

Acknowledgments

The second edition, of course, builds on the first edition. For this reason I would like to reiterate my thanks to those who helped at that stage, including Jocelyn Rashleigh for tidying up my first draft and for a lot of subsequent typing; Alexandra Jefferson, for editorial work, design, typesetting, graphics, patience and production; Bruce Petty for the cartoons; and Construction Information Systems (CIS) General Manager Terry Wright for pushing the project and for his inimitable editorial input and advice.

I must also thank those who kindly and constructively reviewed the first draft of that first edition, including Peter Clarke, Robin Drogemuller, Mark Irving, Tony Mussen, John Rayner and John Tyrril.

CIS staff have again assisted in getting this second edition ready for publication, notably the editor Deborah Singerman who has done a much more professional job of it than I did the first time around. Jocelyn and Terry have once again made their contributions, aided by former NATSPEC editor Greg Turner, current editor John Horridge and NATSPEC manager Noel Burke, who have dealt with my queries and, along with John Schooling and John Rayner, provided valuable feedback on the draft. The views expressed are my own, drawing in part on my own experiences and on conversations over the years with fellow professionals, and are not necessarily those of NATSPEC. And I am of course responsible for errors and remaining anachronisms.

This is the place to note the passing of two people who reviewed the book: John Schooling, the specifier at Devine Erby Mazlin (DEM) and for Bligh Lobb on Stadium Australia, and Peter MacCallum, who for many years was the lecturer in specifications at Sydney University, and active on the Royal Australian Institute of Architects' Practice Committee. I wish well his legacy, the Peter MacCallum Prize, which will be given by the RAIA to a piece of student work each year on, among other things, specification writing.

I am also grateful to the following organisations for permission to reproduce material:

- Australian Building Codes Board.
- Architects Accreditation Council of Australia.
- Construction Publications.
- Department of Public Works and Services, NSW.
- Law Book Company.
- Master Builders Australia.

- National Public Works Council (now Australian Procurement and Construction Council).
- Royal Australian Institute of Architects.
- Standards Australia.
- Construction Information Systems Australia.

John Gelder

Acronyms

AACA	Architects Accreditation Council of Australia
ABCB	Australian Building Codes Board
ACA	Australian Communications Authority
AIA	American Institute of Architects
AIQS	Australian Institute of Quantity Surveyors
ANSI	American National Standards Institute
APAS	Australian Paint Approval Scheme
APCC	Australian Procurement and Construction Council
ASTM	American Society for Testing and Materials
BCA	Building Code of Australia
BREEAM	Building Research Establishment Environmental Assessment Method
BSI	British Standards Institute
CAWS	Common Arrangement of Work Sections
CIL	Construction Information Limited
CIS	Construction Information Systems
CSI	Construction Specifications Institute
CSIRO	Commonwealth Scientific and Industrial Research Organisation
DPWS	Department of Public Works and Services, NSW
FSC	Forest Stewardship Council
GPC	Government Paint Committee
ICIS	International Construction Information Society
IEC	International Electrotechnical Commission
ISO	International Organization for Standardization
JAS-ANZ	Joint Accreditation System of Australia and New Zealand

JCC	Joint Contracts Committee
MBA	Master Builders Australia
NBS	National Building Specification
NES	National Engineering Specification
NIST	National Institute of Standards and Technology
PCA	Property Council of Australia
QAS	Quality Assurance Services
RAIA	Royal Australian Institute of Architects
RIBA	Royal Institute of British Architects
SA	Standards Australia
SSL	Scientific Services Laboratory

1 Historical context

Many models (or simulations) of the unbuilt building are created during the procurement process. One of them is the descriptive contract documentation, prepared to describe the requirements of the purchaser at the nexus, or handover, between building purchaser (principal) and building supplier (contractor).

Descriptive contract documentation has generally evolved from the very simple to the very complex, occasionally going backwards as once-available resources such as literacy have become scarce, notably in the Dark Ages. With the re-emergence of design-and-construct procurement, which dominated the medieval period, it has in some respects nearly gone full circle.

A brief survey of the history of contract documentation follows. A series of topics is examined, beginning with procurement. The various descriptive modes are then explored. Historically, these have been oral description (not discussed here), written material (specifications), drawings and physical models. The scope and nature of their use has varied from culture to culture.[1] They have usually been used in combination.

The emphasis is on written descriptive contract documentation i.e. the specification, and not on other aspects of contract documentation, and certainly not on the buildings and their designers. Appendix 1 provides several parallel example specifications. We see, with the help of a certain amount of conjecture, how the place of the specification in the scheme of things has not been a constant, as the use of other modes of description waxed and waned. In doing so we realise that the current approach to contract documentation is not the only one. There are other possibilities, some radically different. This should open the reader's mind to the changes to current practice suggested later in this book. However, readers keen to get to the business of preparing specifications could skip this chapter.

1. For more on physical models and the impact of information technology, see Gelder (1996a).

Procurement

In ancient Greece the construction industry was quite sophisticated. What we would now call multiple-prime contracting (i.e. with the work carried out by multiple contractors rather than subcontractors) seems to have been common e.g. for the construction of the Erechtheion each part was assigned to a different contractor (Boersma 1970). The Prostoon at Eleusis was built using at least five contracts, in which the work of the stone mason was broken down into various supply and installation packages (Jeppesen 1958). But many projects were single-prime.

This mix of multiple-prime and single-prime contracting, the choice of which depended on the size and nature of the project, continued through to the Middle Ages (for a discussion of modern contracting practice, see Chapter 5). For royal works in the Middle Ages, in contrast to other works, usually no contract sum was stipulated. Given that major works would span wars and changes in the Crown, as well as shifts in architectural fashion and functional requirements, this was only realistic. They would cost what they would cost. Accordingly, the manager of the financial arrangements, the master-of-the-works (the magister), had to be someone trusted by the king to handle large sums of money (and of necessity be both literate and numerate).

This had two interesting consequences: one was the opportunity for embezzlement (one of the masters of the works of King's College Chapel was so charged), and the other was the consequential keeping of careful accounts for every penny spent. These have provided historians with an invaluable record of construction that under lump-sum tendering would not otherwise be available. Indeed the medieval master-of-the-works was much like a modern project manager, with design mainly carried out by master masons, and construction by others still (carpenters, masons and the like).

St Paul's Cathedral (1675-1709) was the first English cathedral to be completed to one person's design and in his lifetime, an aspect of major projects that we take for granted today. Briggs (1927) has Wren "making alterations to his plans and specifications to meet emergencies as they arose", much like a few modern architects.

But there is more to it than this. Wren was fast-tracking, with the first contracts let before the final design was complete. While this echoes the construction of medieval cathedrals, in their case this arose because the design and construction process was often spread across centuries. St Paul's was built in a mere 30 years, and this could only have been achieved by fast-tracking. Besides, the design could not be finalised in advance. It was too radical and there were too many unknowns, to an extent that the design "remained fluid until the moment of actual execution". Also, the 1675 warrant (the commissioning document) had recommended a design that could be finished, and therefore used, in stages (e.g. the choir was opened in 1697). So Wren was responsible for the project management, and developed a rolling program for the works. He was under the control of a Commission, which was responsible for financial and legal matters, the control of the quarries at Portland, and problems of broken contracts (Downes 1987).

The contracts for St Paul's were let directly to separate trades.[2] The first contracts were for work up to ground level, the next for the rusticated ashlar of the basement walls above ground, including windows and arches, and so on. Sometimes two or more contractors per trade were employed:

Thus the first contracts for masonry were made with two separate master-masons ... the line of demarcation between their contracts being fixed at the centre of a window or door opening (Briggs 1927).

Similarly, on the Bank of England (begun in 1788) the architect Sir John Soane dealt directly with contractors in various trades – there was no general 'builder'. But this was soon to change. In Britain the population explosion of 1800-1830 led to a building boom which, in turn, led to the practice of general contracting.

This replaced trade contracting by master craftsmen, which had simply taken too much of the architect's time, both in tendering and managing on-site starts and finishes.[3] Lump-sum competitive tendering by general contractors, who essentially took over the construction management role of the architect, was a more efficient process. An early British example of a general contractor was Thomas Cubitt,[4] active in London in the early 1800s. General contracting appeared in the US before the Civil War, although the term 'general contractor' did not gain currency there until late in the century (Bishir et al 1990). Presumably Australia would have shifted to general contracting at about the same time, say mid-century.

General contracting leads, of course, to the separation of the designer and the executor (subcontractor) of the works. This has consequences for documentation, with clarity, precision and lack of ambiguity becoming much more critical.

Standalone written, and other (e.g. drawn), descriptions were required even for local work, with the general contractor functioning as an intermediary between the designer and the trades. The descriptive documentation had to compensate for this loss of contact by being more explicit. It was also separated from the contract, which addressed the general builder, so that it could be passed onto subcontractors; previously, specifications were generally integrated with the contract (Kostof 1977). Whereas previously the documentation had only addressed one trade per contract, the separation of the descriptive documentation from the contract resulted in multi-trade specifications with chapters to facilitate subcontracting. These, in turn, resulted in a standard preamble for all trades – the 'preliminaries' – to save repetition.

General contracting also forced complete resolution of the project pre-tender, transferring the responsibility for detailing from tradesmen to the documentation team. The consequent increased reliance on the written specification, rather than on memory, permitted and even encouraged a decline in craft skills.

2. *Wren stated in 1681 that "working by Great" – contracting for the whole building – was one of the three principal "ways of working", along with "by the Day", and "by Measure" (Bishir et al 1990).*

3. *Bear in mind that most work was carried out on-site. Prefabrication at the beginning of the century was reserved for mass-produced items such as bricks and tiles. Project-specific off-site fabrication was largely unknown.*

4. *Cubitt established, in 1815, a general building firm employing all building trades on a permanent wage basis. This was the first of its kind. Cubitt was London's greatest speculative builder and developer.*

The building booms of the late 20th century have seen a return to multiple-prime contracting in Australia and elsewhere, and a move away from the general contractor, with its own in-house tradespeople, towards a managing contractor. Given that much modern construction is carried out off-site e.g. prefabrication of lifts, windows, curtain walls, cast stone, this makes a lot of sense.

Trade sections or worksections can be let in a sequence to suit off-site lead times, fast-tracking the construction process and so, in principle at least, keeping the owner's costs down by minimising avoidable (i.e. off-site) delays. Before prefabrication, there would have been little to be gained in this approach. Again, under pressure of time, design-and-construct procurement has become more popular. This enables fast-tracking of both the design and construction processes, further reducing the time from inception to completion of a project.

Written descriptions

In ancient Egypt, only priests could read the hieroglyphics and subsequent hieratic script, and so only they had access to information recorded in these ways. The later demotic script (a simplified form of hieratic) could be read by some in the working classes but was not used much in documentation. For example, the setting-out annotation on the papyrus plan-elevation for the tomb of Ramses IV is written in hieratic rather than demotic script.[5] Priests were essential for its (oral) interpretation.

In spite of problems with literacy the written word thrived in Egypt. Scribes were widespread and used by all classes for formal and informal correspondence. Many late Egyptian papyrus inscriptions (in Greek) survive, and many of these are contracts, accounts, estimates and specifications, including several for civil works (see Appendix 1), one for mosaic tiling (Pollitt 1965) and another for house alterations, all c. 250 BC. Rules for the central registration of contracts at Thebes, from 146 BC, also survive (Hunt and Edgar 1977).

In classical Greece, codes, contracts, specifications and accounts for public works were often inscribed on stone, as part of the public record (wood and papyrus were used for working documents). Many stone inscriptions survive from the 5th and 4th centuries BC.

Perhaps the earliest reference to specifications is to those drawn up by the Greek architect Kallikrates for the temple of Athena Nike on the Athenian Acropolis (in the 5th century BC). Surviving specifications include those for the 4th century BC Prostoon at Eleusis and the arsenal at Piraeus (see Appendix 1) (Jeppesen 1958), from 3rd century BC Lebadeia, and an account for a heavy transport cart at Eleusis (327 BC) (Burford 1969).

Burford (1969) analyses eight contracts (five of which included descriptive material) from all over Greece. Common topics included inspection, responsibility for publication, and fines for damage and bad workmanship. She notes that complete contracts comprised

5. This is parallel to the use of Latin by the Catholic church well into this century.

two parts, the specifications and the rules of play, and that the written descriptions were so detailed they provided adequate instructions for experienced workmen, "needing little or no further amplification (except orally, by the architect, and by demonstrations from drawings or models of the finer details)".

No Roman building contracts or specifications survive (MacDonald 1982). Nevertheless both were used, according to Vitruvius. He refers to contracts at several points, saying that they are to safeguard the interests of both employer and contractor. Other comments indicate that they included the contract sum and date of construction and provided for employer-supplied materials (Morgan 1960).

We can infer from Vitruvius that specifications, separate from contracts, were used.[6] Much of his book is about the correct selection and application of materials, for example, and this is reasonable subject matter for specifications (see Appendix 1). He also advocated Greek practices, and we have seen that the Greeks made good use of specifications. Standards of sorts existed. For example, Vitruvius mentions those for bricks and for plumbing pipes (Morgan 1960).[7] Construction of legionary fortresses relied on generic documentation or paradigms.

For the period between Rome and the Middle Ages, it is estimated that in the year 1000 only 1% of the population of Europe was literate (McEvedy 1972). This did not stop the regulation of craftsmen. Byzantine regulations from c. 900 required craftsmen, if not provided with materials by the client, to give verbal or written notice to the client. Conversely if the craftsmen abandoned the work, the client was to remind them of the verbal or written contract. Client-initiated changes in scope, variations, and poor definition of the nature of the work[8] were covered in these regulations, as were the penalties for structural failure – the contractor was to be flogged, tonsured and have his goods confiscated (Mango 1972).

And for an ecclesiastical audience, illiteracy was not a major issue. Clunaic monks, for example, would build their own monastery and could make good use of a generic or standard design for Clunaic monasteries, such as that described in the Farfa Consuetudinary, a document of 1043. This Latin 'paradigm' comprised a written description of the built parts of the complex, and parts yet to be built (see Appendix 1). It was followed elsewhere for Clunaic monasteries, sometimes closely, sometimes not (Conant 1979). The cost of preparing such documentation was justified by its broad application.[9]

In the Middle Ages, the cause of universal literacy faced a number of difficulties. For example, in 1229 the Inquisition forbade the reading of the Bible by laypersons. The languages used in writing shifted from Latin to the various local languages. At one period in England (c. 1350-1450) contractors had to cope with documentation which might be in Latin, French or English (Salzman 1952). The multiplicity of dialects did not help either e.g. Florentine was not the same as Mantuan. Perhaps not surprisingly French, rather than

6. The word lex is translated variously as contracts (Morgan 1960) and specifications (Granger 1998). In this book, it is assumed that Vitruvius is referring to contracts. The evidence for specifications in Vitruvius is merely inferred.

7. "The pipes get the names of their sizes from the width of the plates, taken in digits, before they are rolled into tubes. Thus, when a pipe is made from a plate fifty digits in width, it will be called a 'fifty', and so on with the rest." (Morgan 1960).

8. Bad specifications have been around for a very long time!

9. Similar arguments today justify master specification systems such as NATSPEC.

Swedish, was used for a 1287 contract for a French master mason to work on the cathedral of Upsala, Sweden (Frisch 1971).

Thanks to these difficulties, much could be concealed from outsiders, particularly in terms of overall design and setting out, which accordingly fell to the designers themselves. An example is the decision taken by the master masons of Strasbourg, Vienna and Salzburg in 1459, that the method of taking an elevation from a plan was to be a trade secret, restricted to guilds (Gimpel 1988; Knoop and Jones 1967).[10] These sorts of precautions were doomed to failure. A few years later printing (Caxton invented the printing press in 1473) threatened their livelihood and this 'secret' was leaked in a publication by a German architect in 1486 (Waugh 1983; Gimpel 1988). Public dissemination of trade secrets – mortar or plaster mixes, how to recognise the grain of stone, how to design an arch – was a direct consequence of information technology.[11]

For procurement of major works it seems that the written word was essential. Those who could write project-specific descriptions, or have them written, did so (in spite of the invention of the printing press, specifications for the most part remained handwritten until the arrival of the typewriter in the 19th century). Many specifications survive. Salzman (1952) collected 123 from Britain, for example (one is a Latin type or standard form of contract, for a house, from c. 1420).

At their simplest these medieval descriptions included the nature of the work, dimensions, enumeration of features such as doors and windows, amounts to be paid in advance and by instalment, completion date, and penalty for non-completion. A few describe the building in detail, mentioning wall thicknesses, timber sizes, materials to be used, and refer to existing buildings as prototypes, or to drawings. Supply of materials (by builder or owner or shared), bonds and sureties, and the provision of lodges is also sometimes covered. The specified-contracts needed considerable expansion and resolution, by literate people such as masters-of-the-works, before a building could be created. For example, the overseer at Reims Cathedral worked out a system of engraved marks for his 3000 statues, so the masons would know exactly where to place each one.

The Italian Renaissance saw the beginnings of a move towards the language(s) of the people. While a useful standard language for scholars and diplomats, Latin was not used on the street.[12] For example, the Italian writer and architect Leon Battista Alberti's (1404-1472) *De re aedificatoria* of 1450 was first produced in Latin, since the elite was the target audience. But manuscript translations into Florentine quickly appeared, with the first printed Italian version in 1546.

Because the medium (paper), and the scribes, were cheaper, and thanks to increased literacy (mostly due to the availability of printed material), use of a written description could expand. Contractually, and for the provision of dimensions, it seems that the written description dominated. For example, the contract of January 1518 regarding

10. This decision was made even though Vitruvius had described how to take an elevation from a plan which, in turn, meant that master masons would know how to do this, and would have recorded the method in lodge books (Gimpel 1988; Stevens 1990).

11. This is a bit like anonymous publication of copyright material on the Internet.

12. This is a similar to today's high-level computer programming language C++, or the hieroglyphics and hieratic scripts of ancient Egypt, or 19th century diplomatic French.

Michelangelo's facade of San Lorenzo in Florence, gives dimensions that differ from, and apparently override, those of the model (Ackerman 1986).

By 1500, about 5% of adult males in north Europe were literate. In 1515 the printing of books was banned unless permitted by the Catholic authorities. The Reformation dealt a blow to these attempts to restrict the flow of information. Rebel Bibles were published in modern European languages, with the result that by 1715 adult male literacy levels in north Europe had risen to 50% or more (McEvedy 1972).[13] The Reformation also ended the monopoly the clergy and merchants had had on writing, which would have kept contracts and specifications as short as possible (except perhaps for large works).

In 16th century France 'controllers' travelled to carry out site inspections and check the works against the contracts – a sign that design export was underway. The specification of 1528 for the royal château of Fontainebleau survives. The masonry and brickwork worksection is 25 pages long and contains around 16,500 words. It provides a complete description of the building, including dimensions such as room sizes and wall thicknesses (implying that these were not on the drawings). It also describes staircases, chimneys, materials and details. The château was built using multiple contracts (Briggs 1927).

At La Muette, Paris (1541-8), the work was carried out "as specified and as shown in the portrait" (Briggs 1927). In 1556 the master mason at Chambord was paid for preparing drawings and specifications for carpentry and masonry (and for superintending the work). Indeed Briggs asserts that a "detailed specification formed the basis of every building contract" (even though elsewhere he simply states that, in a 1608 contract, the work of the masons was to be executed "according to the rules of art").

The 1598 specified-contract for the second court at St John's College, Cambridge, contains key dimensions such as length and breadth, and heights of walls and floors, wall thicknesses and so on, even though drawings were provided (see next section). Other matters included materials, performance requirements, provisions for payment, and the program.

A "loosely worded" specification was written about 1610 by the Jesuit architect Etienne Martellange. It seems to have been standard practice for Jesuits to prepare designs and specifications and for local people faithfully to follow the essential points, however long the building took to complete. The Jesuit architects do not appear to have superintended the work. The documents had to stand alone (Briggs 1927).

A model, a specification ('devis') and a "schedule of prices" were the basis of the contract for the palace at Versailles (begun in 1624). This was usual for Crown contracts. Specifications appear again for the Louvre. Once Gianlorenzo Bernini's designs of 1665 were accepted an "elaborate specification" was prepared and the foundation stone laid. Only then was Bernini's design rejected in favour of Claude Perrault's (Briggs 1927) – surely not the first time a building has been documented but not built, and certainly not the last!

13. In southern Europe, where the Bible was available only in Latin, literacy levels ran at a mere 25% by 1715 (McEvedy 1972).

The 1669 specified-contract for the Bishop's Hostel at Trinity College, Cambridge, contains a set of key dimensions, even though a decent plan, with a scale bar, is provided (Willis and Clark 1988). Clearly specifiers had not yet got used to the provision of scaled and dimensioned contract drawings.

The problem of repetition, first seen in classical Greek specifications (see Appendix 1), and again in the Bishop's Hostel documentation, often remained unsolved into the 18th century. For example, the specification (c. 1740) for a clay-walled house at Colchester repeats dimensions shown on the attached drawing (McCann 1995).

A series of statutes regulating brick sizes was passed during the 18th century (Cruickshank and Wyld 1975). Project specifications were perhaps left with little to say on prescriptive matters covered in statutes. Their skeletal appearance was therefore to some extent deceptive.[14]

While the Bible drove the initial surge in literacy, the novel and the newspaper, which appeared first in the 17th century, were largely responsible for the second and final surge, bringing the level of adult male literacy in 19th century England to 75%.

Books have also impacted as a vehicle for standardisation, through dissemination of standard contracts, of master specifications, and of standards for products and workmanship. Standardisation of written contract documentation – the last great change – began in earnest in the 19th century (AJ Bicknell & Co 1979; Dundin 1988). Forms of contract were standardised (e.g. the three-page US *Uniform Contract*, published 1888 [Dundin 1988]) principally because of the requirements of lending agencies. Over the years, this has reduced legal costs and production costs, and has resulted (generally) in better contracts.

In the 1840s a New York architect, Alexander Jackson Davis (founder of the American Institute of Architects) introduced, along with a standard approach to fees and drawings, perhaps the first standard printed specification form. This included a standard list of 10 drawings and space for customised descriptions, additional to the standard descriptions (Bishir et al 1990).

Written descriptions, like drawings, became more sophisticated. For example, they increasingly referred to proprietary items, reflecting the impact of industrialisation and mass-production on construction practice. An 1810 amendment to the 1774 *Fires Prevention (Metropolis) Act* (London) even cites a proprietary item, Johns' Patent Tessera (a bitumen-based material in sheet form), as being suitable for covering flat roofs and gutters (Read 1986).[15]

In the most developed countries, literacy levels are now between 90 and 100% (Castello-Cortes 1994). Oral and physical descriptions have been largely eradicated, at least for contractual and later purposes, being replaced by paper-based drawn and written descriptions. The specification may combine project-specific material with generic material such as standards, codes, and proprietary literature, commonly including them by reference

14. For example, George Basevi's (1794-1845) 1833 specification for terrace houses in Fulham Road, London, said: "the party walls to be of the thicknesses permitted by the Building Act for 3rd rate houses" (Worsley 1991). "3rd rate houses" were first defined in the 1667 Act; this indicates how specifications may have relied upon the Act since then.

15. A parallel to this is perhaps the code-oriented product certification systems in place today, and the effective citation of two proprietary termite-control systems in the Australian standard AS 3660.1 (1995).

i.e. the specification refers to them, rather than incorporating them physically. This helps reduce the size of the project description, while ensuring that its requirements are not unduly eccentric (much of this generic material is prepared using a process of consensus) and costly. For generic material, universal literacy is essential.

Construction contracts have long been standardised nationally e.g. AS 4000 began life as AS CA24-1952, and the American Institute of Architects A201 traces its origins to the *Uniform Contract* of 1888. Nevertheless, there are several preferred forms for use in Australia for lump-sum contracting (see Chapter 2). They are generally produced by industry organisations (e.g. representing architects, contractors, building owners) together or unilaterally. Specifications have likewise been standardised nationally, through the development of master specification systems, such as NATSPEC (founded as Australian Building Industry Specifications in 1975) in Australia. Many developed nations have such systems in place today, commonly owned by industry organisations.

Last, technical standards for products and workmanship have been produced since the (very) late 19th century by specialist consensus organisations such as the American Society for Testing and Materials (ASTM), founded in 1898. Standards have really taken off over the past 20 years. Standards Australia (which traces its origin to 1922) has around 6500 in its current catalogue (of which almost two thousand have something to do with construction). The numbers reflect decades of standardisation effort, and industry's recognition of the benefits of standardisation.

Specifications have continued to evolve against a general background of increasing industrial standardisation and specialisation, hallmarks of the industrial (and post-industrial) era. Industrialisation has also brought innumerable new methods, components and materials into what is still a conservative industry, inclined to resist such innovation. Modern specifications are extremely specific, with knowledge embedded in standards and product literature, for all to see. There are no guild mysteries any more. Standardisation has further eroded traditional skills (bricklayers do not know their way around AS 3700, painters do not understand terms like latex and solvent-borne), the problem being that, to keep up, you have to read the technical standards and product literature, rather than simply rely on oral tradition. The standards, then, must be cited in specifications.

These traditional skills have been progressively lost in any case, particularly through the impact of the two World Wars, which helped destroy the traditional knowledge base while generating increased demand for built facilities, forcing further industrialisation[16] and the associated specialisation of both designers and contractors. Erosion of the traditional skill base has in turn forced designers to be ever more explicit in their written and drawn descriptions. Things that could once be left unsaid in specifications now cannot.

Our increasing reliance on the written word is compounded by increasing demands by the courts, fewer traditional technical requirements (as they have expanded beyond simple

16. After World War II it was estimated that Australia would have to build 750,000 houses by 1955. The federal Department of Works and Housing marketed prefabricated steel houses for between £600 and £725. Simple to erect, these houses were not, apparently, very popular (Warner 1995).

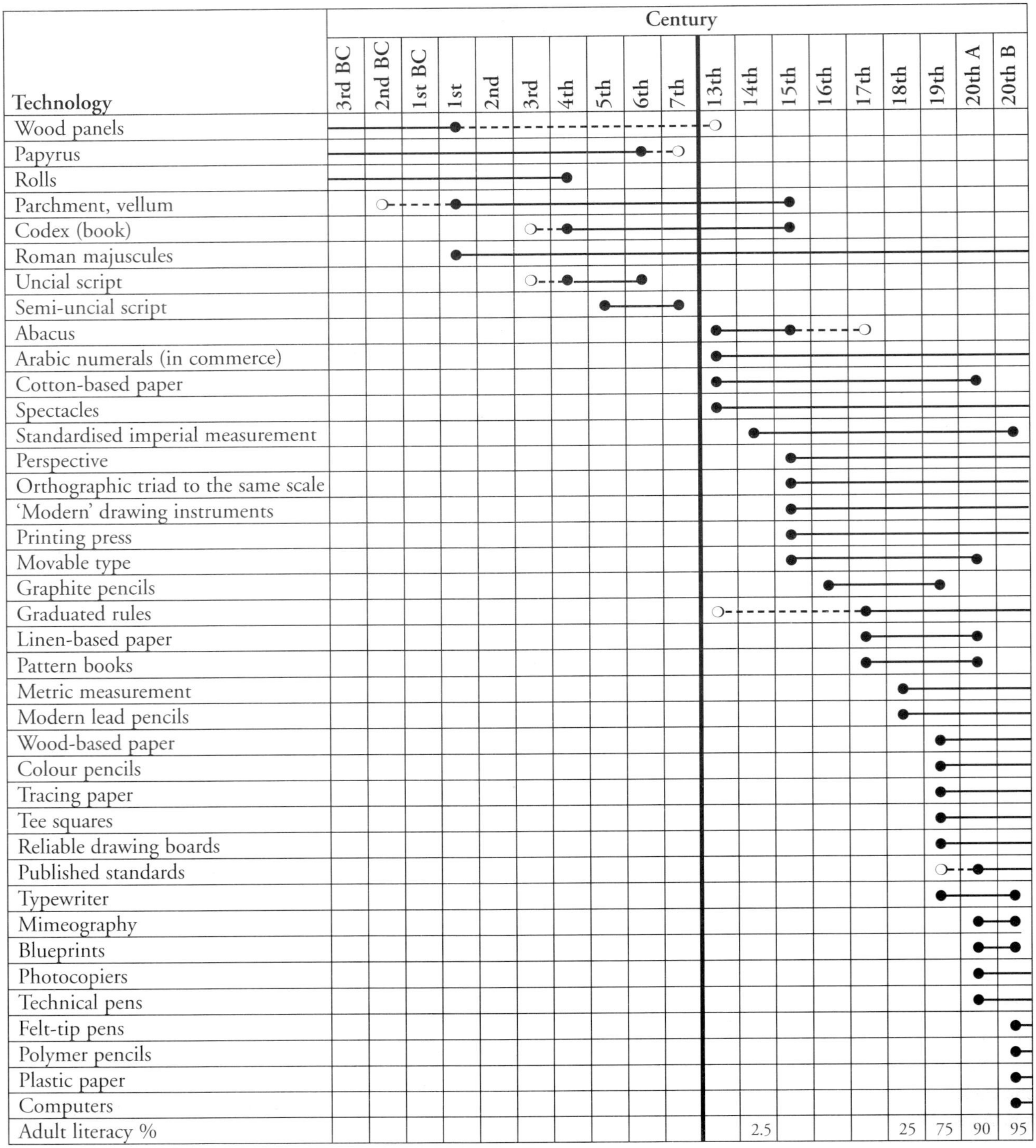

Century
Technology
3rd BC
2nd BC
1st BC
1st
2nd
3rd
4th
5th
6th
7th
13th
14th
15th
16th
17th
18th
19th
20th A
20th B
Wood panels
Papyrus
Rolls
Parchment, vellum
Codex (book)
Roman majuscules
Uncial script
Semi-uncial script
Abacus
Arabic numerals (in commerce)
Cotton-based paper
Spectacles
Standardised imperial measurement
Perspective
Orthographic triad to the same scale
'Modern' drawing instruments
Printing press
Movable type
Graphite pencils
Graduated rules
Linen-based paper
Pattern books
Metric measurement
Modern lead pencils
Wood-based paper
Colour pencils
Tracing paper
Tee squares
Reliable drawing boards
Published standards
Typewriter
Mimeography
Blueprints
Photocopiers
Technical pens
Felt-tip pens
Polymer pencils
Plastic paper
Computers
Adult literacy %
2.5
25
75
90
95

sound construction to encompass a host of other concerns, such as risk to health), incessant product innovation, internationalisation of products, and lately by quality assurance and environmental concerns. Specialist specification writers are the result of these pressures. The ordinary, non-specialist, tradesperson and designer are being left behind.

Drawings

In ancient Egypt, drawings were used in contract documentation, usually annotated. For example, a papyrus plan-elevation survives in fragmentary form, showing the tomb of Ramses IV at Thebes (12th century BC). It is drawn in single-line, with doors in elevation, and is not to scale – the dimensions given in the annotation, and the built tomb itself, do not match the drawn proportions (Clarke and Engelbach 1990; Coulton 1977). So the document was not to be scaled from. A couple of wall reliefs from tombs also combine the plan with elevations of doors and other items, so the use of plan-elevations was evidently quite common. An elevation set out on a grid, for a small carpentry shrine, survives on papyrus. The use of grids was well-known to the Egyptians – the system was used for drawings of human figures, too.

In Greece, full-size drawings of details seem to have been prepared for the temple at the island of Delos, on whitened boards (Coulton 1977). Evidence of their use for other buildings e.g. the 6th century temple of Apollo at Delphi and the 2nd century temple of Athena at Priene in Turkey, hinges on the translation of the words used in surviving inscriptions (Coulton 1977). Architects relied heavily on systems of modules and formulaic conventions tied to the orders (i.e. Doric, Ionic) and their elements. A lot could be left undrawn. Shop drawings also survive, carved on the unfinished inner walls in a passage of the temple of Apollo at Didyma, Turkey (Haselberger 1985).

In Rome, drawings were used by designers. Vitruvius says that the forms of expression of Order are the ground plan, elevation, and perspective. The engineer Frontinus, in his *De aquis urbis Romae* (c. 100), says that plans of aqueducts were always drawn up, so that his office could "deal with any problem immediately, as if it were actually before our eyes and we were deliberating while standing beside it" (Pollitt 1983).

No Roman contract drawings survive, but several record plans do. The marble votive plaque of Claudia Octavia (c. 50) has three plans each at a different scale, and is dimensioned. A mosaic fragment depicting baths has basic room dimensions. On the other hand, the 13 x 18 m Forma Urbis Romae – a marble plan of the city of Rome made between 205 and 208 – is not dimensioned. It is to scale, though prominent temples were shown larger than they really were. Arcades are shown in plan and elevation (Horn and Born 1979).

Table 1.1: Documentation technology. There was little innovation in documentation technology during the 8th to 12th centuries; these were the Dark Ages after all.

Drawings were also used by Roman contractors. Examples include the remains of full-sized working drawings of two pediments, a curved capital and a trapezoid incised into a stone pavement in front of the Mausoleum of Augustus, in Rome. One of the pediments corresponds to that of the Pantheon (completed in 120) (Haselberger 1995).

The Carolingian monastery of St Gall, Switzerland, is recorded in an imperfect freehand "briefly annotated [in Latin] copy" of a plan drawn after the Aachen synods of 816 and 817 (Horn and Born 1979). The plan was drawn to scale, and is modular (850 mm and 13.6 m). It is usually regarded as being a plan of an authorised ideal monastery, which might have served as a type throughout the Carolingian realm – another 'paradigm' (Conant 1979). Proportions, based on sacred numbers, reduced the need for dimensions. Interestingly most of the dimensions given do not agree with the drawing. Rather, they are subsequent corrections, it being far easier to change the dimensions than redraw (Horn and Born 1979). Parts of the St Gall monastery are shown in plan-elevation, the arcaded walls of cloister walks, the arcuated porches of the abbot's house, trees, and crosses.

It has been suggested that architectural drawings were not used before the 13th century, since very few survive from this period. However, it is much more likely that they were used but then discarded, melted down for glue (being drawn on parchment), lost, destroyed to preserve craft secrecy, or to avoid muddling current schemes with abandoned schemes (Briggs 1927). Furthermore, because parchment was expensive people were frugal with it, rarely preparing drawings at all, except for prestigious projects like cathedrals, and perhaps never preparing duplicates. Often only half of a facade or plan was drawn, with the other half being deduced from symmetry (e.g. Reims Palimpset, see below).

As there was only one set of drawings, the designer had to be in control of the project. Documentation management was both easier and more vital in those days. Loss of a drawing would have been catastrophic. It is not easy to imagine working to these sorts of constraints. Reliance on parchment drove the practice of construction.

After the 13th century, contractual elevations and plans were more often drawn. They are referenced in contracts and other literature. Over 100 European examples survive, including the Reims Palimpset (c. 1250), in which one set of drawings was covered by others dated no later than 1270. One of the older drawings is the elevation of a large church, comprising only the left-hand side in detail. Drawings of Strasbourg Cathedral, around 1275 and 1300, and two preliminary designs for Orvieto Cathedral (in Italy), from 1310, survive (Briggs 1927; Millon 1994).

Several lodge books, containing drawings and notes by master masons, also survive. The earliest is de Honnecourt's, but there are at least 10 others from the 14th and 15th centuries. Architects leaving the service of a patron had to leave behind all the drawings they had made in connection with that project (Frisch 1971).

Medieval plans, where used, were different from modern plans. For example, they were rarely dimensioned. Builders relied more on traditional methods and solutions, and easy-to-use ratios, making the plans, perhaps, the least important drawings. Gimpel (Waugh 1983) says that "builders preferred simple ratios ... they were particularly interested in proportions which could easily be translated without recourse to a conversion table".

Full-size details were prepared on-site. At Rosslyn Chapel, Scotland, in 1446 the details were first drawn on whitened wooden boards (echoing Greek practice), which the carpenters carved into patterns or moulds, from which the masons worked. Large shapes such as vaults and arches were inscribed on plaster, and also in the mould lofts or tracing houses.[17] Drawings were executed on canvas, parchment or laths, as "false-moulds" sent to the quarries (Salzman 1952). Masons also engraved full-size details on the structure itself (e.g. the cathedrals at Clermont-Ferrand and Limoges in central France, and at Byland Abbey, England).

Up to the Renaissance the design of buildings changed mostly through evolution rather than, with a few exceptions, revolution. Plans, sections and elevations were barely necessary contractually. Constructors knew what was expected, and details and the odd innovation could be relayed on-site.

The revolution of the Renaissance changed all that. Drawings (and models) were needed to describe what was required. Constructors could no longer rely on past experience. *Disegno* (a sense of design and a skill in drawing) was pretty much the only prerequisite (Alberti added mathematics) for Renaissance architects/engineers who accordingly were often originally jewellers, goldsmiths, sculptors and painters, among whom *disegno* abounded. This led to the separation of the designer from the builder and marked the beginning of the end for the master builder.

Traditional builders (and clients) were out of their depth in this new environment and needed the help that decent drawings could provide. This enhanced the split between the designer and the builder.

The author and architect Antonio Filarete (c. 1460) advocated the need to make drawings (and models) to the same scale. This was another step in the integration of the project description. Large-scale details and a plan of Baldassare Peruzzi's (1481-1536) Palazzo Massini in Rome survive, along with a plan and details for the church of San Domenico at Siena. All these can be seen at the Uffizi Gallery. Both plans are drawn to scale, dimensioned and annotated (Millon 1994).

Modern usage of projection began with Bramante, the architect who picked up Gothic orthogonal and axial traditions, the linkage of the orthographic triad (plan, section and elevation) using the same scale, the use of chiaroscuro on orthogonal elevations, and use of grids of various sizes. His scheme for the dome of St Peter's in Rome is the first in which the plan correlates directly to a section and elevation, and all at the same scale (Frommel 1994). With this type of drawing, architecture could be reproduced, one of the principles of Palladianism (the architectural philosophy of Andrea Palladio and his followers) (Downes 1987).

The architect Raphael initiated the export of design, providing sets of consistently scaled

17. The tracing house at York Minster survives. English documents from 1324 mention tracing houses in the king's works in London (Salzman 1952).

drawings for St Peter's (from 1514), specifically to enable work to proceed on-site during his many absences from Rome (Pugh 1987), incidentally diminishing the role of oral interpretation. This tradition continued into the late 17th century and beyond, with Italian architects supplying drawings or models for buildings abroad, without supervising their erection.

France contributed to the development of projective techniques including other forms of parallel projection such as isometric, axonometric and oblique projections. The 16th century architect Philibert de l'Orme (aka Delorme) extended the use of orthographic projection, to determine the complex shapes needed to construct vaults, arches and stairs, through the construction of *traits* (special drawings for the accurate cutting of complex solids), and generated by a complicated system of projection (Evans 1995).

In England, a 1520 drawing of a tower, perhaps for King's College, Cambridge, survives. The 1598 specified-contract for the second court at St John's College, Cambridge refers to "platts and uprights" (i.e. plans and elevations), which survive, and are all signed and attested (i.e. as part of the contract). They comprise the ground plan, and first and second floor (roof level) plans, and three elevations. All are roughly drawn, to the same scale, with the main dimensions given on the plans. The elevations are colour-coded (Willis and Clark 1988).

From the 17th century, drawings were more widely used thanks to increasing numeracy, to the rapid development of drawing aids, and to the increasing availability of paper. Even reproduction was now commonplace, with prickers used to mark key points of a document through to a blank sheet of paper below (Hambly 1982).

With the widespread use of pattern books – collections of standard designs and details – from the 17th century onwards, particularly for less prestigious projects, there was less need for precise project-specific drawings, which had always been costly to prepare and reproduce. Clients or their designers would select the designs they wanted in their buildings, and craftsmen would reproduce them. Through them, every visually significant part of a building could be fully documented.

The labour involved in the preparation of engravings for printing was justified by the mass market. Everyone could access a wide range of designs and details without having to go to the trouble of preparing them for each project. The use of pattern books, widely regarded as dictated by fashion, was more probably a case of making a virtue out of a practical necessity i.e. to maximise the extent of designed and documented work (and hence to maximise the transfer of the power to design to the client) while keeping the cost within reasonable bounds.

The Robert and John Smythson drawings in the Royal Institute of British Architects library (dated around 1599-1632) include plans drawn to scale, often with the dimensions of the principal rooms. A number are working drawings,[18] but these do not include

18. Working drawings included "A Draughte For the Platte of a rounde window" which had setting-out instructions and a "Scalle of Footes" (Kostoff 1977).

detailed instructions, which were instead often included in the contract along with information on major roof structures (which presumably were custom-designed). That is, the specification was integrated with the contract (Kostof 1977).

The architect Francois Mansart's contract drawing for the Church of the Visitation in Paris, of 1633, is a good example of a type of drawing developed in France for contractual use, showing several complementary views of the entire building – half-elevation, half-section, quarter-plan and quarter internal elevation – all on the same page. Interestingly, the built church varies this drawing considerably (Downes 1987).

Christopher Wren (1632-1723) used drawings extensively for the design and construction of St Paul's Cathedral. About 200 survive out of perhaps 600 that were prepared. Most of the surviving drawings were superseded, discarded or unfinished. Only one set of final drawings was prepared, and few have survived the inevitable damage in use and later disposal. Interestingly, one surviving drawing is of an as-built paving plan, a type of drawing apparently without precedent. Wren did some drafting himself, but employed and trained many assistants to draw and contribute in other ways (Downes 1987).

The drawings for St Paul's were to scale, and many scales were used. Scale bars were sometimes given, and less often the scale ratio (e.g. 5 feet). Non-presentation drawings such as studies were sometimes dimensioned. Annotation is not common (Downes 1987).

After the Renaissance the next step was to discard the revived classical traditions so that, like today, virtually any style became possible. Hence Mannerism, Baroque, Rococo and Neo-classicism, which throughout the mid-1700s rapidly followed each other in various parts of Europe. For this approach to architecture, thorough documentation – both drawn and written – was essential. There was too much innovation, with new theories and changing styles, for builders to rely on tradition.

Paper mills were established in England in the mid 18th century, so paper was cheaper and available in a wider range of sizes. Architects took advantage of this and from about 1790 prepared many more drawings than they had done before. They also used colour wash coding (e.g. black for existing fabric, yellow for timber), which became increasingly standardised among offices though no standard as such was published (Worsley 1991).

Projective geometry had been developed further. The standardised arrangement of plan, front and side elevation known as 'first angle projection' originated in the book *Descriptive Geometry* (1795) by French military engineer Gaspard Monge. Monge's techniques were first applied to the design of fortifications, and so had been a state secret since about 1780. But they became part of the curriculum of the Ecole Polytechnique, France's influential school of architecture, and found their way into Germany, the USA and the UK in the first half of the 19th century via French authors and instructors (Pugh 1987; Stevens 1990).

Ease of reproduction is vital in tendering and contractual use, where many copies are needed. Difficulties in reproduction of drawings (and specifications) hindered, in a practical sense, developments in building procurement. Tendering and subcontracting were difficult, for example, with only one full set of drawings on most projects. This must have been frustrating. They could make lots of copies of specifications and contracts on large state projects, which were tendered using printed documents (tenderers would inspect the drawings in a central location), so they could see that not being able to mass produce contractual drawings was limiting their options.

The introduction in the 19th century of linen paper, tracing paper and carbon paper opened up building procurement to other possibilities, such as subcontracting, as did the introduction of cheap paper made from wood pulp (Lambert 1983). The drawing board, tee-square, set-square and ruling pens were refined during the 19th century under pressure from engineers (a profession which, on the military and civil side, goes back at least to classical Rome). Accuracy and precision were becoming more important, as engineering drafting shifted from pictorial to "more schematic forms of representation which were less ambiguous and capable of more precise interpretation" (Pugh 1987).

With the growth of new industrial technologies during the industrial revolution, working drawings became more complex. An early exemplar of the new technology was the coal-firing of iron for the Coalbrookdale Iron Bridge, England, finished in 1779. The best known example of this increasing complexity, however, is Joseph Paxton's pioneering prefabricated Crystal Palace, completed in 1851. Most of the work was done off-site. This required a completely different approach to the preparation of working drawings e.g. everything had to be detailed to ensure it would fit together and nothing could be left to chance. Prefabrication, however, proved be something of a dead-end at the time.

Nowadays, working drawings have evolved into stylised, standardised creations. Annotation is, or should be, minimal. Dimensions are laid out according to established standards. Accepted scales are standardised too. The emphasis is on speed of creation, clarity, consistency and reproducibility e.g. by reduction photocopying and microfilming. Drafting has become standardised through the use of published standards, defining, for example which colours and types of hatching represent which materials, and which symbols represent which electrical fixtures or weld types. A good understanding of these standards is essential for the correct interpretation of drawings. Standardisation of drafting protocols has become more important with the use of the computer (which has brought its own problems of convention such as layering), and the international trade in drafting software (Hall 1999).

Reproduction has shifted from manual to mechanical so that today the 'original' drawing (or piece of text) lives on a computer's hard disk. In effect, every print is an original.

Models

In ancient Egypt, models may well have been used in building procurement. The Egyptians certainly knew how to make them. Models of buildings exist, but are all funerary. Some are quite sophisticated, for example, two wooden models dating from around 1990 BC, show a walled garden, complete with copper-lined pool, detailed trees, columns, and copper rainspouts, doors with bolts and grilled windows (Tooley 1995).

In classical Greece, full-size models of architectural details are known to have been supplied to contractors e.g. in standards for marble tiles, as at Assos in Turkey (Coulton 1977). Models in wax were provided for rosettes and coffer lid acanthi at the Erechtheum (Athens), and for plasterwork, a metal grille, painted lions' heads and fine mouldings at Epidaurus (Greece) (Burford 1969). Project models of complete buildings are not known.

Dilke (1987) suggests it was more common for Roman architects to make scale models than drawings (Vitruvius, however, seems to suggest the opposite), and their use is well attested (Coulton 1977), at least for promoting designs to prospective clients in competition. But this is at the design end of the architectural process. In documentation, the extent of Roman usage of models is not known.

In Europe it seems that no plaster or wood models were made over the period 1050 to 1350 (Waugh 1983). Certainly north of the Alps models were little used for structures before the 16th century (Salzman says there is no evidence for the use of models in medieval England). In 14th century Italy large-scale models were used by master builders for cathedrals in Milan, Florence and Bologna (1390), as well as for lesser structures (Millon 1994). It has been suggested that these early Italian models "could have been the response of the master builders to the illusionism of the painter-architects" (Frommel 1994).

Medieval master masons travelled widely, often for specific commissions, recording designs and details as they went, for future use elsewhere. Reference buildings were used for major public works but these were large buildings and far afield. Henry III's Westminster Abbey, based on Reims with features from Amiens and Ste Chapelle in Paris, is one example. A number of contracts, even for modest projects, required the use of reference buildings e.g. in England, the roof of the church at Halstead in 1413 was to follow that at Romford.[19] Reference buildings were less parochial in design and construction, and served to educate the local workforce in new techniques.

In Renaissance Italy, models were used extensively for study, presentation, competition and construction. Hundreds, perhaps thousands were built.[20] St Peter's in Rome provides a 100-year history of models, from those by Bramante to those by Carlo Maderno (1556-1629). Brunelleschi, in particular, relied on models. He had to demonstrate through prototypes that his radical construction methods for the massive dome of Florence Cathedral were viable.

19. This is parallel to the use of a reference project in PC-1 (1998) clause 9.1 (see Appendix 3).

20. An exhibition entitled The Renaissance from Brunelleschi to Michelangelo: the representation of architecture, was held originally at the Palazzo Grassi in Venice in 1994 (Millon and Lampugnani 1994).

Alberti rejected models that were "colored and lewdly dressed with the allurement of painting" (Rykwert 1988). Similarly, Brunelleschi built models that were incomplete, and without ornament. This may have been simply to keep the cost of the models within reason, but seems mostly to have been about maintaining central control of the design and construction process by giving people only as much information as they needed at the time.

Alberti strongly recommended the use of models for design development and cost control, and saw the model and orthogonal drawings as complementary, with the drawings serving the other functions e.g. to instruct the builder. But both Brunelleschi and Michelangelo used models specifically to guide construction work. Design models were often modified to become construction models. Purpose-made construction models and models of details were also used, often full size, a good example being Michelangelo's cornice at the Farnese Palace, Rome (Millon 1994).

The later 16th century models were more complete, down to sculptural detail. The wooden model of Antonio da Sangallo the Younger's project for St Peter's, built between 1539 and 1546, is the largest surviving model of the Italian Renaissance, at 4.57 m high, 7.8 m long, 5.79 m wide (Millon 1994).[21]

As contract documents, these later models emphasised certainty. Less is left undecided, less could be altered during construction, and the client had a much better idea of what was on offer. This shift presumably reflected bad experiences with the earlier spartan models, but it also increased the cost of the models, and the time needed to make them. That, and the availability of a more versatile and less costly alternative (i.e. drawings), probably triggered their demise.[22]

Wren's 6 m long wooden Great Model for St Paul's (not the final design)[23] was made in 1673 both to inform the project's benefactors "that comprehend not Designes and Draughts on paper", and also to inform the "inferior Artificers", to give them a clearer understanding of what was involved. Yet the so-called Definitive design was constructed without a model, which was unusual for major projects. This was because Wren "resolved to make no more Models, or publicly expose his Drawings, which (as he had found by Experience) did but lose Time, and subjected his Business Many Times, to incompetent Judges" (Wren's son, quoted in Downes 1987).

Accordingly the project relied to an unprecedented degree on the drawings prepared by Wren's office, and on Wren's oral interpretation. But the drawings were probably not used much 'on the scaffold'. Both joiners and masons made models, moulds and templates for various parts of the building for this purpose (Downes 1987). These would have been based on the drawings. Most of these models ended up being used for firewood in the architect's offices (Briggs 1927)!

The use of models in general seems to have lapsed particularly during the 18th and 19th centuries under Beaux Arts influence. Nevertheless for some 20th century designers (e.g.

21. Sangallo's design was never built. He knew it never would be and so put all his efforts into the model, as a record of his skill for posterity.

22. Wooden models continued to be used in the Baroque period. An exhibition entitled The Triumph of the Baroque: Architecture in Europe 1600-1750 (at Turin, Montreal, Washington and Marseilles, 1999-2001) shows 80 of them (Morris 1999).

23. The models of St Paul's were the subject of a 1991 exhibition at the Royal Academy of Arts in London entitled Sir Christopher Wren and the Making of St. Paul's.

Gaudi; Utzon; Skidmore, Owings & Merrill; Ando; Mitchell/Giurgola & Thorp; Ove Arup & Partners) physical models have remained an important tool in detail design and, especially, documentation.[24]

Summary

This brief history of contract documentation reveals a number of trends. The relative importance and usefulness of the media of descriptive contract documentation or modelling (oral, written, drawn and three-dimensional) have been greatly influenced by available information technology and skills (see Table 1.1). This parallels the influence that construction technologies (such as steel and concrete framing, lifts and airconditioning) and skills have had on the design and construction of buildings.

Information technology (in the general sense) has, over time, worked to remove or circumvent disadvantages embedded in particular media for documentation and has, occasionally, opened up the possibilities beyond simply making documentation in a particular medium easier to prepare and use. For example, Renaissance developments, such as perspective, orthographic projection, and cheap paper, not only made drawings easier to prepare and use, but also enabled architects to replace master builders.

Literacy has been a rare and valued attribute throughout much of history, and has fundamentally affected the modelling of buildings. Literate people can write descriptions of buildings, and can read them. Illiterate people cannot. Until the 19th century (in Europe, North America and Australia), literacy split society in two, with an illiterate mass on the one hand, and a literate elite on the other. Both groups built buildings, but went about it in different ways, the illiterate relying on oral description and tradition, the literate making good use of written descriptions (specifications) and innovation. In these places after the 19th century, this split rapidly disappeared.

The written description has evolved. In the Middle Ages, written descriptions of materials and workmanship were short. They included room dimensions and quantities, and were often combined with contractual matters. Frequently these descriptions were the only documentation, though simple drawings were attached to some contracts. Accordingly, supplementary models and/or drawings were prepared in advance to resolve the design or communicate it to the client. They were also prepared later to resolve construction and detailing, notwithstanding the continued importance of oral interpretation.

This situation continued into the Renaissance after which written descriptions expanded, for the following reasons:

- Literacy became more widespread.
- Technical requirements became more complex.
- Trade skills declined as did the desire for contractor design initiatives.

24. Three full-sized mock-ups are discussed in McCallum et al (1998): a wall component for a medical centre, a criminal courtroom, and an advanced measurement laboratory. They were used to assist in design, documentation and construction, and were seen as being extremely cost-effective (also refer to Morris [1996]).

- Production and reproduction of the specification became easier.
- There was a growing need for documentation to stand on its own.

Offsetting this expansion, material was relocated. First, room dimensions were transferred to the drawings, which were now supplied routinely. For smaller projects, the entire specification was often located on the drawings anyway, minimising the effect of the removal of this information from the specification. Contractual material was separated out, although an outline description is still a part of many building contracts. Quantities too were relocated to separate bills, at the same time, to assist the general contractor.

Nevertheless, the written description has continued to expand, even where 'traditional' construction is required. Partly to contain this trend, the written description relies increasingly on external documents, with an emphasis on standardisation of requirements and formats. Experience has also encouraged the specifier to write documents capable of one interpretation only. Loose (or expansive and rambling) wording is not wanted. Nor is repetition. This is a qualitative shift in specification practice.

Contractual drawings in the Middle Ages were basic where they existed at all – plans and elevations at most, all to different scales. Even pre-contractual and post-contractual drawings were not common, other than full-sized detail templates. Partly this was a function of the available technology. This under-utilisation of drawings meant that design and construction could only evolve slowly.

This changed with the Italian Renaissance, with the preparation of many drawings throughout a project's life. Various newly developed projective techniques were used, notably perspective, the same-scale orthographic triad, and drawings of details. Renaissance designers needed to be able to develop radically new ideas (to serve the new classicism and gunpowder-based militarism). Inexpensive cotton-based paper and new mathematics gave them the means, and enabled designer-driven experimentation with projective techniques.

Since then architecture has never had the same influence on information technology. Rather, architects have used and refined what others have created. Conversely, information technology – and building modelling – has never had a similar effect on architecture. It is reasonable to suggest that the creation of these projective techniques by Renaissance designers in turn both created and sustained the modern architectural profession. Without them, it would not exist.

Before the Renaissance, drawings were not usually dimensioned, but were sometimes set out on grids to facilitate transfer from drawing to facade for carving. Many drawings were not drawn accurately to scale. In any case, few people would have known how to resolve, or what to do with, dimensions or scales. Accordingly, set-out was commonly done directly by the architect. The spread of numeracy and further standardisation of units made it easy and desirable to use dimensions on drawings. With the industrial era, drawings have become even more numerous, more precise, more standardised, and use of dimensions has grown.

Depiction of a three-dimensional object on two-dimensional material has always been problematic. Until it could be done satisfactorily, physical models or precursor buildings i.e. an existing building somewhere, often had to be used to convey the design concept accurately. Before the Renaissance, models were mainly used post-contractually, as full-size moulds for details. Whole-building models have only ever been used contractually as a matter of routine in the Italian Renaissance. After that they reverted to a presentation role, as drawings were found to be a great deal more convenient to prepare and use. Large-scale models were extremely expensive to build, they took a long time to make, were not portable, and were often out of date even before they were finished.

2 Contemporary context

This chapter looks at the context of the contemporary Australian specification, comprising the construction industry generally, the project process, contract documentation, and last, as a small but important part of a designer's vocation. The relationship between the specification and various contextual elements is briefly explored. Specifications, being pivotal in contract documentation, an understanding of their place in the scheme of things is essential. A parallel chapter on context in the UK and USA follows, for readers who might find themselves working in those places, who seek to compare the Australian scene with that overseas, or who are just curious. Readers familiar with the contemporary context could go directly to Chapter 4, where discussion of how to produce a specification begins.

A number of elements of the Australian construction industry impact on the work of specifiers. Excluding construction materials and technology (not the subject of this book) the more significant elements discussed throughout the chapter are as follows:

- Regulatory requirements.
- Standards and conformance system.
- Quality assurance.
- Tendering practice.
- Conditions of contract.
- Master specifications.
- Classification systems.
- Environmental building assessment schemes.
- Project process.
- Vocation and training.

Regulatory requirements

Australia uses a two-volume building code, the Building Code of Australia (BCA). Currently there are state and territory variations to the national code dealing with specific regional or local situations, though some of these, such as earthquake, cyclone and alpine construction, are dealt with in the national code. Other regional differences are progressively being incorporated into the national code with the goal of eliminating all regional supplementary requirements. The BCA is published for a central authority (the Australian Building Codes Board [ABCB]) and the regional (i.e. state and territory) variations are included as appendices.

The BCA was republished in 1996 as a mainly performance-based document with a view to eventual harmonisation with the New Zealand code (the New Zealand and British codes are also performance-based). Like the New Zealand and British codes, it relies heavily on deemed-to-satisfy text and referenced standards. The BCA has been adapted for use in a number of Pacific Island nations and was used in the development of the national construction law for Vietnam (Wallis 1995). Volume Two, specific to Class 1 and Class 10 buildings (i.e. housing), is known as the *Housing Provisions*.

Project specifications (and drawings) must indicate proposed methods of compliance with the BCA and other statutory requirements, including any BCA options adopted, particularly deemed-to-satisfy standards, such as AS 3700 (1998) for masonry. Because the performance-based code relies substantially on such standards, so will project specifications.

The ABCB has a product certification scheme for product compliance with the BCA in the absence of suitable standards. The Commonwealth Scientific and Industrial Research Organisation (CSIRO) operates a building products and systems appraisals program, replacing Australian Building Systems Appraisal Council (ABSAC) Technical Opinions. These are prepared where no standards exist but, unlike the ABCB scheme, acceptance of CSIRO Appraisals by local authorities is not mandatory.

Statutory requirements for building services are generally covered elsewhere, and are usually administered by agencies other than the local authorities (local councils) which administer the BCA. Australian Communications Authority (ACA) standards deal with telecommunications, for example, and are administered by the ACA under the national *Telecommunications Act 1997*. Plumbing requirements are covered by AS/NZS 3500 (various) and are generally administered by the various water supply authorities, which often impose supplementary requirements, under state legislation. However, there are moves to integrate AS/NZS 3500 with the BCA (AS/NZS 3500.3.2 and AS/NZS 3500.5 are already cited in the BCA), and so its various parts have been split into performance requirements and prescriptive solutions.

Electrical requirements are covered by AS/NZS 3000 (wiring rules) and administered by the various power supply authorities under state legislation. These, too, are being split into prescriptive and performance-based parts. Domestic versions of AS/NZS 3500 (AS/NZS 3500.5 [2000]) and AS/NZS 3000 (AS/NZS 3018 [1997]) again match the BCA. Gas installations are based on the Australian Gas Association (AGA) code AG 601 (now AS 5601) and are administered by local gas supply authorities under state legislation. State-based occupational health and safety legislation also affects many aspects of construction, from boiler manufacture to roofing installation.[1]

1. Safety on site, normally seen as the contractor's affair, is actually part of the architect's business in Britain (under the elaborate Construction (Design and Management) Regulations 1994 [see Joyce 1995]), in South Australia, under Schedule 5 Clause 1(c) in the South Australian Development Regulations 1993, in Western Australia (see RAIA Advisory Note AN14.01.612 [1998d]) and in New Zealand under the Building Code at Section F5 Construction and demolition hazards.

Standards and conformance system

The principal standards organisation in Australia is Standards Australia. Its standards are designated AS (for Australian Standard), followed by a four digit number e.g. AS 1214. The date and title (e.g. 1983 *Hot-dip galvanized coatings on threaded fasteners [ISO metric coarse thread series]*) may also be stated.

Increasingly, Australian standards are repackaged international standards i.e. those produced by the International Organization for Standardization (ISO) and the International Electrotechnical Commission (IEC). Similarly, joint Australian/New Zealand standards are becoming more common (50% of all new projects undertaken by Standards Australia), due to the general desire to facilitate trade under the Australia-New Zealand Closer Economic Relations Agreement (SA 2000).

Standards are consensus documents (the committees that prepare them represent a range of interests including government, manufacturers, users and academia) and so may not fully reflect the needs of purchasers (for whom specifiers act). Accordingly, purchaser-operated schemes exist which extend standards. One such scheme is run by the Australian Paint Approval Scheme (APAS), whose GPC (Government Paint Committee) specifications expand and tighten the relevant paint standards.

In the absence of Australian standards, pragmatism prevails. A number of foreign standards are in use, particularly British Standards Institute (BSI) and various American standards, including ones produced by the American Society for Testing and Materials (ASTM) and the American Architectural Manufacturers Association (AAMA).

Standards are widely used in the construction sector, in regulatory roles (e.g. via the BCA), contractually (e.g. via the specification), and voluntarily (the quality assurance standards are a good example). Indeed, specifiers cannot avoid them, as Figure 2.1 illustrates.

Standards Australia is trying to assist specifiers by publishing supplements and guides and by splitting major standards e.g. into housing, design, construction, testing, and maintenance. Summaries and extracts of some are available as handbooks. Hard copy and downloadable (.pdf) versions of Australian and international standards are available from the Standards Australia website.[2] Drafts and amendments can be downloaded gratis and the site has a number of customer services to make it easier to keep track of changes to standards.

Standards are also published in a range of relatively inexpensive hard copy packages (notably the Collection series), and on CD.

It is all very well having product standards, but whose products conform to them (if any)? Manufacturers (first parties) may make claims of conformance, and a number of purchasers (second parties) assess conformance for themselves. But more reliable than the first, and less expensive than the second, are independent (third-party) product certification schemes. These are the recommended sources of information on conformance to standards. They are important for regulatory approval purposes, where the standards are cited in legislation, and are potentially very useful for specifiers and contractors who need a convenient means of demonstrating compliance with specified standards.[3]

2. See www.standards.com.au.

3. For specifiers, Australia's standards and conformance infrastructure mostly means Standards Australia, the National Association of Testing Authorities Australia (NATA) and the Joint Accreditation System of Australia and New Zealand (JAS-ANZ) (DIST 1995).

The StandardsMark scheme (e.g for security screen doors), operated by Quality Assurance Services (QAS)[4], is the best-known product certification scheme. In the building sector it covers around 30 standards (including parts of standards), and has just over 120 licensees, with over half these standards attracting only one licensee. About a sixth of the licensees are foreign (QAS 2000). The registers for plumbing and electrical products are on the QAS website.

There are a number of other schemes. The Plywood Association of Australia (PAA) operates its Quality Control and Product Certification Scheme nationally. Pine Australia (PA) has a Quality Certification Scheme. The Glued Laminated Timber Association of Australia (GLTAA), the Carpet Institute of Australia (CIA), the Australasian Window Council (AWC), the Australian Timber Industry Certification (ATIC) and the Australian Window Association (AWA) all operate schemes in their subject areas.

The Woolmark and Woolmark Blend schemes (both operated by The Woolmark Company)[5] and the Australasian Furnishing Research and Development Institute (AFRDI) Blue Tick are other examples. APAS runs a major scheme for paints, which certifies against the GPC specifications (see above), and the associated Scientific Services Laboratory (SSL) runs another for active fire protection equipment (1999), which includes recognition of overseas product certification (see Gelder 1997a).

Regulatory product certification schemes include the *WaterMark* and *Plumbing Safety TypeTest Mark*, the *Electrical Safety TypeTest Mark*, AGA/ALPGA (Australian Gas Association/Australian Liquefied Petroleum Gas Association) *Approved Appliances*, ACA (Australian Communications Authority) *Certified Components*, the *Water Conservation Rating Scheme*, the *Energy Labelling Scheme*, and the *Electromagnetic Compatibility Framework* and *Regulatory Compliance Marks* (Gelder 1997a).

Third-party confirmation of a manufacturer's claims does not always rely on the existence of a standard. We have noted that the ABCB certification scheme assesses products against BCA requirements where there is no relevant standard – the BCA is the 'standard' here. CSIRO Appraisals, Building Research Association of New Zealand (BRANZ) Appraisals, and British Board of Agrément Certificates are assessed against the manufacturer's own claims for the product or system, not against a published standard.

The Joint Accreditation System of Australia and New Zealand (JAS-ANZ)[6] is responsible for the accreditation of product certification bodies (PAA was among the first to have been accredited) and for accrediting agencies for environmental certification of products to ecolabelling standards in the ISO 14000 series.

Quality assurance

Put simply, quality assurance (developed originally in Britain in the form of BS 5750 – quality systems) is about stating what you do by documenting processes in a procedures manual, doing it by following the procedures, and proving that you have done it by undergoing regular audits of procedures.[7]

4. *See www.qas.com.au.*
5. *See www.wool.com.au.*
6. *See www.jas-anz.com.au.*
7. *Quality assurance should not be confused with total quality management (rarely seen in the Australian construction sector). On the differences and links between quality control, quality assurance and total quality management, see Gelder 1992.*

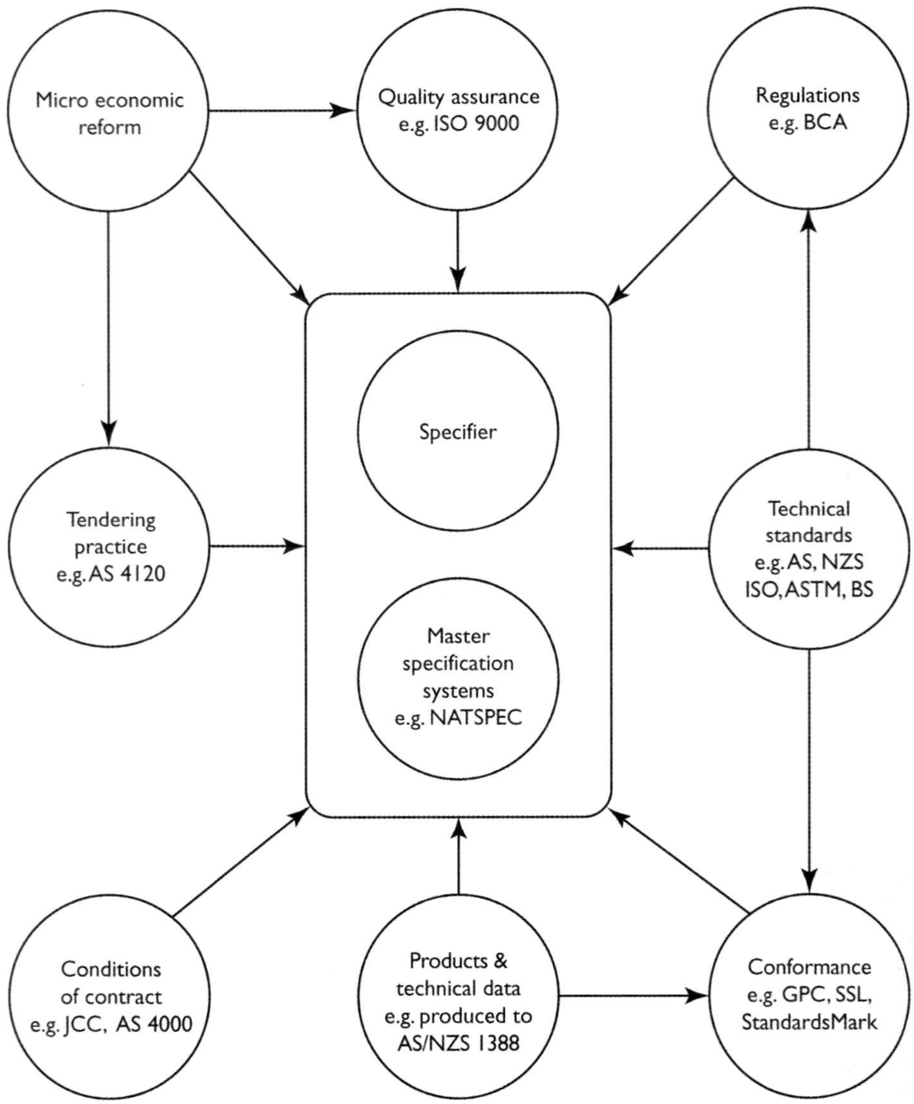

Figure 2.1: The specifier in the construction industry framework. Almost everywhere the specifier looks, there are standards of one sort or another with which familiarity is recommended, if not actually essential. Arrows indicate direction of influence or input.

Quality assurance is about the processes of the quality management system, not the products that are produced. It is intended to assure the customer that the supplier has taken all reasonable steps to provide consistent products and services. In other words, where principals require quality assurance of consultants and contractors, it is in part because they no longer wish to rely on the word or reputation of these suppliers for quality management, and/or because they wish to reduce their direct vetting of the products supplied. In short, they want the suppliers to be fully responsible for their own work, in line with the Abrahamson risk allocation principles outlined in Chapter 5, and they want proof that this responsibility is taken seriously.

Australia has a well-advanced quality assurance program for construction, based on the use of the AS/NZS ISO 9000 series of quality management standards.

There are over 20 quality system certification bodies in Australia, accredited under the JAS-ANZ scheme, which can certify companies to the quality assurance standards and undertake regular audits to verify compliance with documented procedures. These bodies must operate independently of those consultancies involved in quality assurance training, documentation and related activities.

Many Australian construction sector manufacturers and suppliers are certified to the quality assurance standards, as are quite a few contractors. Certified consultants are becoming more common (around 300 architectural practices)[8], although they are under pressure from public works authorities that see quality assurance as a substitute for traditional quality control procedures.[9] In these procedures, the principal reviews the contract documentation prepared by consultants and the contract administrator inspects and tests building work carried out by the contractor.[10]

So the widespread view that quality assurance (of processes) is a substitute for quality control (of products) is misleading, even faulty. To quote *GPC News*, July 1995:

Quality assurance provides a level of assurance that a product will achieve the manufacturer's required standard of conformance, but not necessarily of a quality or indeed a standard of conformance that will satisfy the purchaser.

Here is a good summary of the state of confusion, from *Pine News*, March 1996, quoting the US National Research Council:

Awareness of quality system registration has expanded rapidly in recent years, in conjunction with global growth in demand for the ISO 9000 series of standards. ... There is a danger that much of this demand is due to misunderstanding of the ISO 9000 series link to product quality – in some instances through over promotion by the burgeoning service industry of quality systems auditors, registrars, trainers and consultants. Despite its rising popularity in European consumer markets, it was not intended, and is not appropriate, as a certificate of product quality for use in the consumer marketplace.

Conformance to the specification and drawings must be separately demonstrated, using quality control. Mere confidence based on QA is not enough. It is only a starting point in

8. This number includes firms where each state office is certified and also includes non-architectural organisations where only one section is capable of doing architectural work (JAS-ANZ 1999).

9. An example of a QA system for small architectural offices is Nelson's ABC2 Architects Model Quality Management System (1998).

10. The definitive source for QA-certified organisations is The JAS-ANZ Register of Accredited and Certified Organizations, published and maintained by Standards Australia.

contractor (or manufacturer, or consultant) selection. This point is made with respect to manufactured products in Gelder (1997a). An important difference for the products and services of consultants and contractors is that project-specific quality manuals will be prepared by the suppliers that will be sighted by the principal/purchaser, who can then decide if the proposed quality management systems are adequate. This access is not available to purchasers of manufactured products (see also Equity in Chapter 5).

With respect to contractors, the contract administrator is the neutral third party nominated in the general conditions to administer the contract on a day-to-day basis.[11] One of the contract administrator's functions is to ensure that both parties meet their obligations. In the case of the contractor this has traditionally meant, through periodic 'observation', checking that the built product complies with the specification and drawings – checking, and certifying, conformance. The parallels with third-party product certification should be clear.[12]

Under quality assurance, the product-checking role of the contract administrator does not change, because the contractor's quality assurance does not guarantee conformance of the product to the specification, only conformance of the procedures to the quality procedures manual. Nevertheless, part of this product checking may be formally delegated to the contractor, and will then be reviewed as a process by the quality assurance auditor. Auditor training and certification is clearly a vital component of the implementation of quality assurance, and is coordinated by JAS-ANZ and carried out by the Quality Society of Australasia. Special audit procedures are needed for project-based activities, which are quite distinct from those of manufacturing.

Quality assurance has a fundamental impact on the content and style of specifications, through hold and witness points, inspection and test points,[13] and the contract administrator's less interventionist role. This last point means that the contractor may be expected to sign-off more on conformance with the specification and drawings (the contractor's inspection and test plan is one of the audited documents). Requirements for products must be expressed more clearly in the specification than we are perhaps accustomed to. The documents are even more standalone under quality assurance. The contractor must be absolutely clear as to what level of quality is required, otherwise it cannot honestly claim conformance, and so it will be more inclined to take the documents literally.

Traditional forms of contract (e.g. SBW-2, JCC, AS 4000) are based on quality control rather than quality assurance, i.e. they assume appropriate 'observation' and certification by the principal (or the principal's agent). It could be argued that the optional quality assurance clause (29.2) in AS 4000 is an inadequate way of dealing with the matter. The role of the contract administrator and the responsibilities of the contractor and principal under a quality assurance regime are, or should be, fundamentally different to those under a quality control regime.[14] For JCC a similar clause was considered, but not adopted.

11. In AS 4000 (1997), the term used for the contract administrator is "superintendent", in SBW-2 and JCC it is "architect". C21 cuts out the middleman and uses "principal". It also refers to a "principal's representative".

12. A parallel scientific approach to building certification would be beneficial e.g. sampling, Pareto analysis (see Chapter 8).

*13. **Hold point:** work subject to a hold point must not proceed beyond that point until the contract administrator has seen it; reasonable notice must be given to the contract administrator.*
***Witness point:** work subject to a witness point can proceed beyond that point unseen by the contract administrator, but the contract administrator must be given reasonable notice.*
***Inspection and test point:** work subject to an inspection and test point must demonstrably pass inspection or test before proceeding beyond that point.*

14. For example, Cornick (1991) argues that considerations of quality management would have the specifier spelling out construction methods (means) as well as results (ends). This is at odds with sensible practice (see Chapter 7).

Tendering practice and conditions of contract

The Construction Industry Development Agency (CIDA) prepared a model tendering code, initially published as an interim standard, but now available as AS 4120 (1994). This was preceded by the *No Dispute Paper 2A Tender Code and Code of Ethics*. Victoria and Western Australia use AS 4120 and the DPWS (NSW) has prepared its own tendering code which is actually quite close to AS 4120. The standard addresses the issue of ethics in particular, and is widely supported in the private sector.

No Dispute, the joint NPWC/National Building and Construction Council (NPWC/NBCC) 1990 report on minimising construction sector claims and disputes and improving work practices and the quality of construction, states at recommendation 1.5:

Standard Form Contracts should be adhered to where possible and, when amended, amendments should be clearly identified.

One might expect, accordingly, a move towards fewer standard forms of contract, but this is unfortunately being undermined by the actions of a few key players, producing competing non-consensus contracts. We are extremely unlikely to attain the Dutch position of a single national lump-sum contract, for both the private and public sector. Nor will we even achieve the American position of two dominant contracts. Several consensus[15] forms are available, at the time of writing, including the following:

- AS 4000 (1997) is used in civil engineering, and has been adopted for use by some members of the APCC. South Australia and Queensland will continue to use its predecessor, AS 2124 (1992). The contract is intended for major projects, and has an associated form of subcontract (AS 4901 [1998]). It supersedes AS 2124, which is still available.
- The JCC series, published by the Joint Contracts Committee which comprises the Royal Australian Institute of Architects (RAIA), Master Builders Australia (MBA) and the Property Council of Australia (PCA). The four current versions (C, D, E and F) were published in 1994. The contract is intended for major projects, and has an associated form of subcontract (SC-JCC).
- SBW-2, prepared jointly by the RAIA and MBA. The RAIA has prepared special conditions and local versions to suit various applications around the country e.g. Victoria and Queensland (see for example RAIA Advisory Note AN15.02.102 [2000] – proprietor in occupation). This is the RAIA's most popular form of contract. It is intended for simple works projects. The MBA publishes an associated form of subcontract (SBWSC2).
- The NSW Government's C21 contract, used by the DPWS and other departments for major works. It was first published in 1996. The 1999 edition is a consensus form, having benefited from a consultative process.

Non-consensus, often unilateral, forms of contract include the following:

- The PCA's *Project Contract* PC-1, for major works, published in 1998. The associated

15. Consensus forms are prepared by two or more of principal, contractor and contract administrator representative organisations.

form of subcontract is PSC-1 1998.

- The *New Works and Services Contract* (Allen Allen & Hemsley 1994), for major works.
- The Department of Defence suite of contracts, 1992, which deal with, among other things, design-and-construct and document-and-construct procurement.
- The *Construction Industry Contract* CIC-1, AN15.11.100 (1998e), developed by the RAIA for use in the construction industry.[16]
- The suite of contracts for residential construction published by the NSW Department of Fair Trading, including *Minor Works Building Contract* (1998b), for projects under $5000, *Renovations Contract* (1998c), for projects between $5000 and $25,000, *Home Building Contract* (1998a), for projects under $25,000 and the *Swimming Pool Contract* (1998d).
- RAIA or MBA conditions of contract such as MWC-1 (minor works), ABP-1 (administration by principal, aka Standards Australia miscellaneous publication MP65) (by RAIA); and AAD1 (minor works), BC3 (commercial works), and PM 2 (project management) (by MBA).
- Public sector minor works general conditions of contract e.g. DPWS (NSW) (1999b).
- Those produced by major employers e.g. Joint House Department of Parliament House, Canberra.

New forms of contract are required for recent developments in the industrial environment such as quality assurance and total quality management, partnering and alliancing, design-and-construct contracting, document-and-construct contracting, build-own-operate-and-transfer (BOOT) procurement, turnkey procurement[17], pre-qualification criteria and performance documentation.

Standards Australia has produced a design-and-construct contract, AS 4300 (1995). For partnering, it has been suggested that the British *New Engineering Contract* 1994 may be "a preferable vehicle" (Tyrril 1995). The specifications will be different, too. For BOOT projects, the handover documentation may need to specify many items not normally considered, such as qualifications for staffing personnel, linen and other loose items, stationary design, and operating procedures. Some of these would also feature in turn-key specifications.

Master specifications

NATSPEC is the only peer-reviewed national master building specification system in Australia. It is published by Construction Information Systems (CIS)[18] Australia which has 20 shareholders, including the RAIA, MBA, Institution of Engineers Australia (IEAust), Australian Institute of Quantity Surveyors (AIQS) and individual APCC members. Shareholders all have the opportunity to contribute to NATSPEC worksections (see also Chapter 4). Most APCC members require the use of NATSPEC by their consultants.

NATSPEC is closely aligned with the BCA (the six-monthly updates are aligned with the latest BCA Amendment) and relies on the use of Australian and other reference standards.

16. At time of publishing CIC-1 was replaced by Australian Building Industry Contracts (ABIC) MW-1 2001 major works contract, the first in a suite of contracts published jointly by the RAIA and the MBA.

17. Turnkey projects are those which are handed over to the client complete and ready to operate (i.e. fully furnished including consumables) – all that is needed is to turn the key.

18. See www.cis.com.au.

NATSPEC material includes supplementary descriptive and performance specifications, prompts for specifiers to complete and options for them to select. The specifier may specify proprietary products if so desired, and can override defaults such as standards. While many standards cited reflect regulatory requirements, many other standards are referenced and are useful in particular where principals do not allow specification by proprietary item, or where 'essential characteristics' of specified proprietary items need to be defined to ease the substitutions process. This is discussed more fully in Chapter 6.

There are a number of NATSPEC specification packages: *BASIC* is for simple building projects where brevity is a priority; *BUILDING* is for more complex projects, and is NATSPEC's definitive specification; *SITE+STRUCTURE* is for civil engineering and landscaping work; *SERVICES* covers hydraulic, fire, electrical and mechanical installations; and *DOMESTIC* is for individual houses. Each package has a *Commentary* to help specifiers customise and supplement the NATSPEC material for specific projects. All packages are available on CD, along with other relevant material such as the BCA.

The system currently works by deselection of inapplicable material from the default one-part master material, rather than by selection from a library of alternative clauses. That is, it is subtractive rather than additive. NATSPEC worksection structure is discussed in Chapter 4.

The specification packages are available only on subscription; a print version of *NATSPEC DOMESTIC – the better homes specification* is also available.

Classification systems

Different types of construction entities e.g. worksections, products, elements, parts, and spaces, need different classification systems. A single system for all entities is not possible. It is difficult enough getting to a single system for a single type of entity. This does not help the specifier in structuring the specification, working with material from various sources or organising the library.

NATSPEC is organised around a draft (1998) worksection classification system, *NATSPEC Classification 2*, devised in a dialogue with the Association for Co-ordinated Building Information in New Zealand (ACBINZ),[19] and inspired by the UK Common Arrangement of Work Sections (CAWS) system (see Chapter 3) (Wright 1998a).

Product information in Australia is organised by publisher-specific systems, for example, Info-Link uses one system and Selector.com uses another system. Other classification systems are in use overseas (see Chapter 3 for examples).

Environmental building assessment schemes

There are presently no environmental building assessment schemes in Australia but any future scheme will be an important driver for product selection and specification.

The situation is changing all the time and parts of a building environmental assessment scheme do exist. There are several operational energy schemes for buildings, the most

19. ACBINZ has since been taken over by CIS' New Zealand equivalent Construction Information Limited (CIL).

important being *NatHERS (National House Energy Rating Scheme)*, distributed by CSIRO through Unisearch.[20]

The NSW Sustainable Energy Development Authority (SEDA)[21] has a number of energy efficiency schemes notably the Energy Smart Homes program, an Energy Smart Government program for government agencies, and a greenhouse rating scheme for commercial buildings.

Several schemes consider various aspects of green materials selection, including the *BES (Building Material Ecological Sustainability) Index* developed by Partridge Partners (NSW), the CSIRO's prototype *Embodied Energy Module* (unpublished), and the *EcoSpecifier Handbook*,[22] developed by the RMIT Centre for Design, EcoRecycle Victoria and the Society for Responsible Design.

Project process

This book assumes that design consultants and others follow a standard project process (see Figure 2.2), which includes the production and use of the specification. The project process relates to the life-cycle of the building project, at least in so far as it involves the architect and other designers. To avoid misunderstanding, the project process and its relationship to the project's life-cycle are spelt out below (based on RAIA Advisory Note AN10.01.101 [1993a]).

Essentially a design consultant's activities fall into eight phases, linked to the four-step project sequence outline in the Preface, as follows:

- Inception: Pre-design.
  Site analysis.
- Design: Schematic design.
  Design development.
  Contract documentation.
- Execution: Tendering and negotiation.
  Contract administration – construction.
- Finish: Contract administration – post-construction.

Each of these general phases can of course be broken down into others (see Figure 2.2). The RAIA breaks its eight fee-oriented phases into core services as follows[23] (this is an expansion of the outline given above):

- Pre-design: (Preparation of brief.)
  Consultation with authorities.
  Preliminary feasibility study.
  Form of building contract.
  Selection and engagement of consultants.
- Site analysis: Consultation with authorities.
  (Existing buildings survey.)

20. The statutory position with respect to the environment is set to change, with the ABCB (1999) stating that, "A review of BCA objectives … needs to include consideration of maintenance, durability, energy efficiency and environmental outcomes". Energy efficiency is likely to be the first cab off the rank with mandatory minimum energy performance requirements being developed for incorporation in the BCA (SPECnews 2000).

21. See www.seda.nsw.gov.au.

22. See ecospecifier.rmit.edu.au.

*23. Selected non-core services are in brackets. Services pertinent to specification preparation and use are in **bold**.*

Site analysis.
Detailed off-site services studies.
(Environmental studies and reports.)

- Schematic design:
  Consultation with authorities.
  Architectural design.
  Opinion of probable cost and program.
  (Preliminary economic feasibility study.)
  Presentation to principal.
  (**Development approval**)[24]
  (including materials and finishes schedules).
- Design development:
  Consultation with authorities.
  Architectural design.
  Opinion of probable cost and program.
  Presentation to principal.
  (**Development approval**)
  (including materials and finishes schedules).
- Contract documentation:
  Consultation with authorities.
  Architectural documentation (including specification).
  (**Special materials research.**)
  (**Special tender documents/scheduling.**)
  Opinion of probable cost and program.
  (**Building approval.**)[25]
- Tendering and negotiation:
  Tender process.
  Document revisions (including specification).
  (Negotiated contracts.)
  Contract preparation and execution.
- Contract administration – construction:
  Consultation with authorities.
  Supplementary details (and specification).
  Contract administration.
- Contract administration – post-construction:
  (**Record drawings**) (and specification).
  (**Review of warranties and guarantees.**)
  Defects liability period.
  Final certificate.

24. 'Development approval' is known by a different name in each state , as follows:
- New South Wales: development consent.
- Australian Capital Territory: Design and siting approval.
- Queensland: Development permit for development.
- Victoria: Town planning approval.
- South Australia: Development approval.
- Western Australia: Development approval.
- Tasmania: Planning approval.
- Northern Territory: Development consent.

25. 'Building approval' is known by a different name in each state, as follows:
- New South Wales: Construction certificate.
- Australian Capital Territory: Building approval.
- Queensland: Development permit for building work.
- Victoria: Building permit.
- South Australia: Development approval.
- Western Australia: Building licence.
- Tasmania: Building permit.
- Northern Territory: Building permit.

The development approval application is normally submitted at the end of the schematic design phase. Design development and documentation often proceed without approval having been received – a sometimes risky proposition (though it is nice to have a keen client). Clearly in these circumstances it is sensible to be as sure as possible of approval before lodging the plans with the local authority. The building approval application is normally submitted early in the contract documentation phase. The project may be put on hold at this point, pending the outcome of the local authority's deliberations. More often,

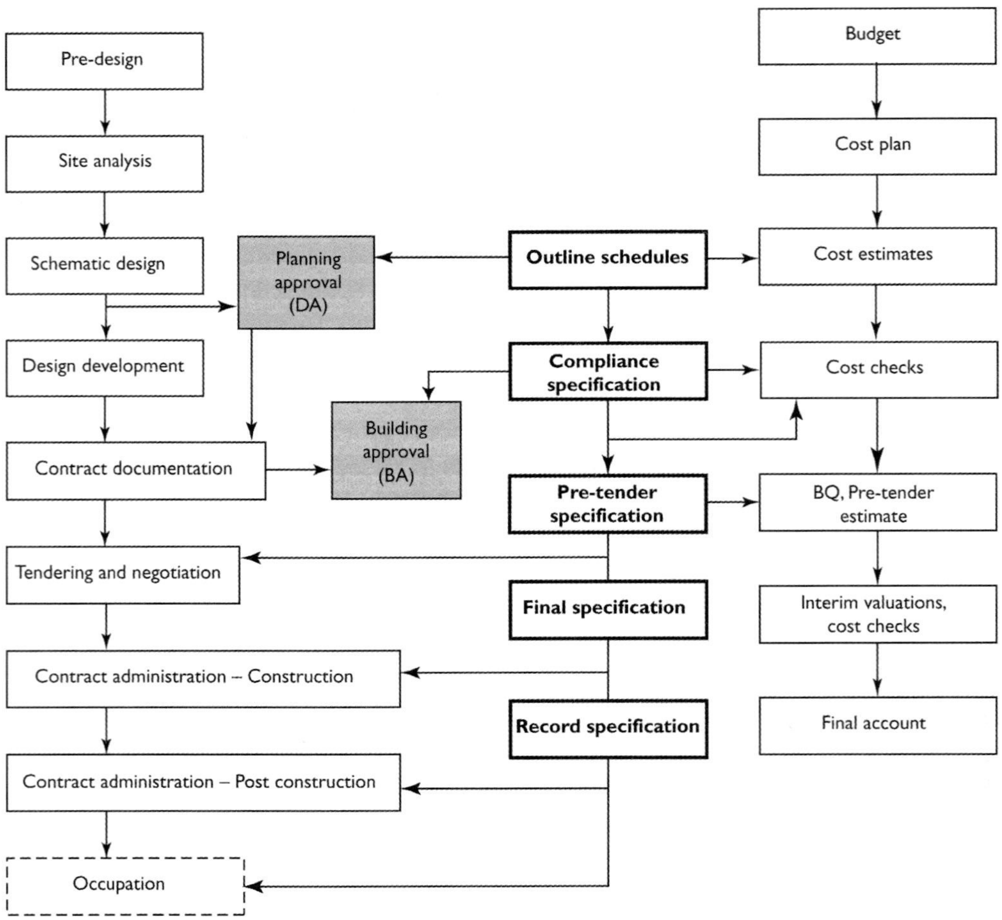

Figure 2.2: Project process

documentation proceeds, the client being prepared to accept the risk of outright rejection and the high holding costs of the site making time short.

The specification begins life during the schematic design phase, as the materials and finishes schedules that are part of the development approval application. Generally, most of the specification is prepared during contract documentation, and it is used in all stages thereafter. An embryonic version (say, 20% complete) is submitted as part of the building approval application (this is discussed more fully in Chapter 8).

In Australia (unlike Britain and South-East Asian countries such as Singapore) it is not usual for quantity surveyors either to write the specification or to prepare specified bills. They are, of course, involved in the vital business of cost planning, for which the specification is an important source of information. The principal specifier is usually the architect. Engineering and other material is provided by the appropriate specialist consultants i.e. structural and civil engineers, services engineers, landscape architects and the like.

Table 2.1 summarises the elements of contract documentation in Australia and identifies the national source material (often Australian standards). The specification is pivotal as it provides a link to the conditions of contract, through the Preliminaries, and to the drawings, through cross-references in the worksections.

Vocation and training

The Architects Accreditation Council of Australia (AACA) has produced *National Competency Standards in Architecture*,[26] which are intended to define the extent and depth of a practising architect's competencies, initially as a basis for assessing the qualifications of overseas-trained architects wishing to practice here. The *Standards* include material on specifications, which can also readily be applied to non-architect specifiers.

The *Standards* consist of four units: design, documentation, project management and practice management. Each unit is subdivided, creating a series of 13 contexts, which are in turn divided into a total of 39 elements against which are set a series of performance criteria (see Appendix 2). Figure 2.3 shows the units, and relevant contexts and elements. Only one of these 39 elements particularly addresses specifications.[27]

Under the construction documentation context, the four elements, each addressed to a degree in this book, are:

- Establish a documentation process.
- Prepare architectural drawings with regard to the location, extent of building elements, components, finishes, fittings and systems.
- Prepare architectural specifications and schedules.
- Coordinate the project documentation.

26. The National Competency Standards were adopted by the AACA in September 1993, and are under review.

27. This means, perhaps, that there could be 38 books this size covering each of the other elements! One might also expect the subject to be given 1/39th of total course time at tertiary institutions. For a five-year full-time architecture course this would equate to five full weeks on the subject.

Element	Function	Source
Tendering documentation	Instructions to would-be constructors i.e. tenderers	AS 4120 (originally produced by CIDA). NATSPEC offers supplementary material in its *Tendering* worksection
General conditions of contract (GCC – different forms offer different coverage)	Essentially the rules of play for the principal and contractor, and must be read with project specific material, located in the Appendix for SBW-2 and JCC, and Annexure Part A for AS 4000. They attempt to deal fairly and thoroughly with 'what if' scenarios, as well as routine administration matters	AS 4000, JCC C & D, SBW-2. Others are available, including C21, PC-1, the British *New Engineering Contract*, and AS 4300 for design-and-construct work. Some principals have their own
Special conditions of contract	Project specific variations to the GCC. These are located in Annexure Part B for remote editing of AS 4000, and Section 15, for JCC and SBW-2	Advice produced by APCC and RAIA, for example. Some principals have their own
Specifications – Preliminaries	Addressed to the contractor and comprising additional commercial material, response to GCC prompts for proactive decisions (material which otherwise has no effect on GCC), project administration requirements and use of infrastructure. Does not include contractor's preliminaries, probably not principal's policies, and not special conditions	NATSPEC *Preliminaries* (for each contract). Some principals have their own e.g. public works authorities
Specification – General requirements	Governing all worksections i.e. common material of interest to trades and subcontractors	*General requirements* (in *NATSPEC BASIC, BUILDING* and *SERVICES*). Commonly partly integrated with the *Preliminaries*, and partly with the worksections
Specifications – Trades or technical material, also known as worksections	Written description of the works, addressed to the contractor, but in practical terms used by the subcontractors and suppliers	NATSPEC worksections (*BASIC, BUILDING* or *SERVICES*)
Specifications – Multi-trade schedules	Multi-trade descriptions of the works e.g. finishes, doors	NATSPEC *Multi-trade schedules*, in the *Templates*
Drawings	Graphical description of the works. The key drawings are the plans, which are generally multi-trade. Details tend to be more aligned with particular worksections in the specification. Graphical schedules may be provided	AS 1100 (various) – these standards are the bibles for engineers, but are not much used by architects, for whom favourite texts serve as defaults. Offices may use their own standard details. Books of standard details and project details are available e.g. from the RAIA and Architectural Press
Bills of quantities	Numerical description of the work, using quantities e.g. m^2, units, L. Derived from drawings and the specification. Bills may be contractual, not contractual, or not provided. Generally used only for larger projects e.g. over $2 million. Bills are not much used in civil work or services. Combined specification and bills are rare	*Australian Standard Method of Measurement* (ASMM) Edition 5 (AIQS 1990). Standard short forms are available e.g. DPWS (NSW) *Concise Method of Measurement for Construction Works* (1993)

Table 2.1: Contract documentation elements – functions and sources

Specifications are clearly a part of an architect's (and engineer's) job and skills. We may quibble with the detail of the *Performance criteria*. For example, specifications do not define the demarcation of separate trades and subcontractors, cross-referencing between other documents and the specification is not always helpful, and performance should only be specified in certain circumstances. Nevertheless, as we will see later, these *Standards* delineate clear expectations for professionally produced specifications.

A number of architects run specialised specifications consultancies, offering a project specification preparation service, an office master preparation and maintenance service, project document review, assistance during contract administration, and so on (Heineman 1999a; 1999b). This type of consultancy is much more common in the USA than here, though several such practices exist throughout Australia.

The business of preparing specifications is taught (with limited contact hours) at undergraduate level in most, if not all, Australian schools of architecture.[28] Building undergraduates are taught the subject at some universities but as far as the author knows, engineers, interior designers, quantity surveyors, facility managers, project managers and landscape architects are not taught the subject at all. Nor are there any postgraduate courses in contract documentation generally, let alone specification writing in particular, which is surprising given the amount of time spent on this activity in practice.[29]

28. In 1997 Sydney University provided 14 contact hours on the subject, and the University of Technology Sydney provided 10.

29. For architects, between 1 and 6% of fees (2.5 to 15% of 40% – see Chapter 4) are earned by preparation of specifications. Over a five-year course this equates, if course time is allocated in proportion to fees earned from a subject, to 2 to 12 weeks of course time. So whatever way we look at it, undergraduates are grossly underexposed to specifications. And there is a lot to learn.

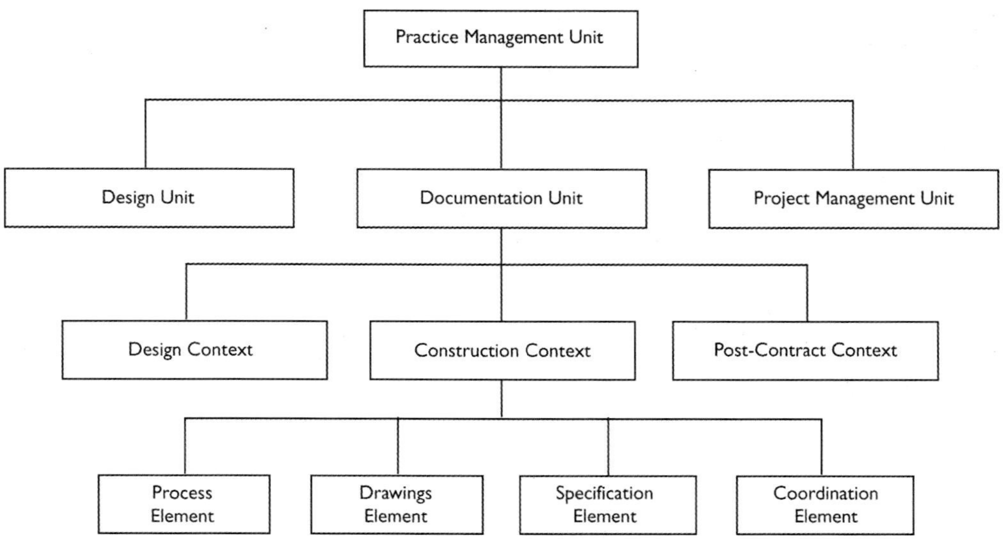

Figure 2.3: Specification preparation as a part of vocation/competency (AACA 1993). Competency in specifications is explicitly required.

No wonder specifications are often neither done well nor taken seriously. Many construction sector documentors are incapable of critically reviewing the specifications prepared in their offices – they have no 'theory' against which to test them.

Texts available include this book, of course, and Standen's (1995) and Kalms' (1990), plus texts prepared overseas (the general principles of specifying are the same), so teachers are not without resources. For practising architects and other designers, public lectures and workshops on the subject are offered occasionally.

Summary

For specifiers the contemporary construction industry is characterised by an increasing use of standards. For a start, they are cited in legislation governing the industry, such as the BCA, where they are cited as deemed-to-satisfy solutions. This role of standards is likely to increase with the shift to performance-based legislation.

Standards are also the basis of quality assurance, tendering practice, many conditions of contract and subcontract (not all published by Standards Australia, but industry standards nonetheless) and, of course, master specification systems. The use and nature of standards is discussed more fully in later chapters.

Specification production and usage is an integral part of the construction project process. Many people contribute to their production and many people use them, including the authorities for approval purposes, and estimators (commonly quantity surveyors) for cost control purposes. The specification may be seen as a pivotal part of construction documentation, particularly as it has links to the conditions of contract, standards, manufacturer's literature and the drawings.

The specification is, therefore, an essential part of an architect's or engineer's vocation and so should not be viewed as something to be left till the end of the documentation, and used only as a last resort. Its production must be properly resourced – this will be discussed in Chapter 4 – and the document must be used, like the drawings, as a matter of routine.

Finally, given the constantly changing context for building projects and the industry generally, specifiers would be well advised to review traditional approaches to their task with an open mind, especially as many of these approaches may no longer be valid.

3 Contemporary context — United Kingdom and United States

This chapter parallels the previous one on Australia by looking at the context of contemporary United Kingdom and United States specification practice, comprising the construction industry generally, the project process, and contract documentation. The interest of context in these countries is that it derives from the same historical (British) model as Australia, so similarities and differences are illuminating for contemporary Australian specifying practice. Some Australian specifiers will work in these countries or under their systems (e.g. to an extent Singapore follows UK practice and South Korea, USA practice), so the chapter has practical benefit too.[1]

As for Australia, a number of elements of the UK and US construction industries impact on the work of specifiers. Significant elements discussed throughout the chapter are as follows:

- Regulatory requirements.
- Standards and conformance system.
- Master specifications.
- Classification systems.
- Metrication.
- Environmental building assessment schemes.
- Project process.
- Vocation and training.

Regulatory requirements

In the UK, the performance-based *Building Regulations 1991* (the current edition) were first introduced in 1985. They apply in England and Wales. There are different versions for Scotland and Northern Ireland. The performance-based regulations are served by a series of prescriptive Approved Documents, as follows:

1. *For specification practice in other countries, and more on context and practice in the UK and US, see McGregor (1997).*

- A Structure (1992, amended 1994).
- B Fire safety (2000).
- C Site preparation and resistance to moisture (1992).
- D Toxic substances (1985, amended 1992).
- E Resistance to the passage of sound (1992).
- F Ventilation (1995).
- G Hygiene (1992).
- H Drainage and waste disposal (1989, amended 1992).
- J Heat producing appliances (1989, amended 1992).
- K Protection from falling, collision and impact (1998).
- L Conservation of fuel and power (1995).
- M Access and facilities for disabled people (1999).
- N Glazing – Safety in relation to impact, opening and cleaning (1998).
- Reg 7 Materials and workmanship (1999).
- Timber intermediate floors for dwellings (excluding compartment floors) (1992).
- Basements for dwellings (1997).

The Approved Documents are available as individual booklets, as a set, and on CD. They are drafted as recommendations rather than as deemed-to-satisfy specifications (e.g. using "should" rather than 'shall' or 'must', as in "Turf and other vegetable matter should be removed from the ground to be covered by the building at least to a depth sufficient to prevent later growth" from Approved Document C clause 1.2). This means they need to be rewritten into project specifications rather than simply cross-referenced as can be done with the BCA prescriptive provisions.

At the moment the US construction sector is marked by a total lack of uniformity in building codes. Only 31 of 50 states have statewide codes. The remainder rely on city or district codes, or even no codes at all, as in large unincorporated parts of Texas (Bullen 1991).

There are three model codes – the National, the Standard, and the Uniform – but they are not applied anywhere in their current editions (Bullen 1991). The situation can become extremely complex. Kansas City, for example, has 44 jurisdictions. Of these, 35 use the UBC (but different editions) and the rest the NBC (four different editions, beginning with 1975). All three model codes are used in the regulated parts of Texas.

The model code agencies are private sector organisations (secretariats) which maintain the model codes but have little influence on the use or misuse of their material by the local building authorities. A number of problems arise from this, including the widespread use of different editions. This leads to 20-30% of cited standards being out-of-date – a major problem given the frequency of revision and editorial change of American standards (sometimes more than once a year; ASTM C140-99b [testing masonry units] was published three times in 1999, for example).

Each code has at least five parts (building, plumbing, mechanical, fire prevention, existing structures), intended to be self-contained to suit their different audiences. This has inevitably led to overlapping and conflicting information e.g. fire-related aspects are dealt with in three parts. Authorities occasionally mix parts from different codes, necessitating major amendments to bring the different parts into line with each other.

Authority-specific editions are unique to those authorities, just as the old Australian Uniform Building Regulations Coordinating Council (AUBRCC)-based codes used to be in Australia. A central national standard part, and a set of authority variations attached as in BCA96, would just be too cumbersome (but, revealing, making the extent of unwarranted inconsistency between jurisdictions clear).

Because the codes are prescriptive, if a type of construction is not regulated it is effectively banned. Special dispensation must be sought to use it. Accordingly a few jurisdictions have developed codes for specific types of construction such as straw bale, adobe and rammed earth e.g. Austin, Boulder, New Mexico, Pima County and Tucson, and California. Others have not.[2]

Performance-based codes (which would simplify the problems with particular types of construction) are not planned generally for the US in the near future (Vognild 1998). The model codes, however, have been harmonised, with a private sector agency, the International (so-called) Code Council (ICC)[3], and one series of codes, the International Codes, replacing the three separate agencies and codes.

On top of the codes is a wide range of national regulations that may override local codes. These are published in the *Code of Federal Regulations*. Entities responsible, and the topics they deal with, are as follows (Bullen 1991):

- Housing and Urban Development (HUD): manufactured housing, energy.
- National Institute of Standards and Technology (NIST): building technology, fire research.
- Occupational Safety and Health Administration (OSHA): construction site safety.
- Department of Energy (DOE): building energy efficiency, energy conservation, and solar.
- Architectural and Transportation Barriers Compliance Board (ATBCB): accessibility standards e.g. *Americans with Disabilities Act* (ADA).
- Environmental Protection Agency (EPA): air quality, and toxic substances e.g. National Emission Standards for Hazardous Air Pollutants.
- Federal Emergency Management Agency (FEMA): building seismic safety.

Standards and conformance system

The British Standards Institute (BSI)[4] is the principal UK standards-producing agency. While many British Standards (BSs) are still in use, and are cited in the Building

2. For more information, refer to the following websites: The Earth Building Foundation at www.earthbuilding.com; and US Department of Energy Centre of Excellence for Sustainable Development at www.sustainable.doe.gov.

3. See www.intlcode.org.

4. See www.bsi-global.com.

Regulations and master specification systems, increasingly the UK is adopting EuroNorms, as BS ENs, and ISO and IEC documents, as BS EN ISOs and BS EN IECs. BSI is active in European and international standards agencies, so quite a few of these standards have their origin in British Standards. Perhaps the best known examples are ISO 9001, for quality management systems (originally BS 5750), and ISO 14001, for environmental management systems (originally BS 7750).

On conformance, BSI Product Services operates the well-known Kitemark scheme, equivalent to the StandardsMark scheme. Kitemark schemes exist for hundreds of standards, including ISO and IEC standards, and even one Norwegian Standard. Building products covered include, for example, ready-mixed concrete (companies whose cement is certified are listed on the BSI website), construction sealants, fibreboards, electronic ballasts, concrete tiles, fire extinguishers, sanitary tapware, and door closers. Certification involves testing of the product or service against the performance requirements of the relevant standard, and an assessment of the manufacturer's quality management systems against the ISO 9000 series.

BSI Product Services is also a Notified Body for many European Union (EU) Directives, and so can function as an independent certifier under the European CE mark scheme. CE marks indicate that the manufacturer claims the product conforms to the minimum requirements for health and safety laid down in EU Directives flowing from the EU Construction Products Directive.

The government-backed British Board of Agrément (BBA)[5] is a member of the European Union of Agrément (UEAtc), and operates an Approved Products scheme. Agrément Certificates cover durability, installation and compliance with the Building Regulations. BBA is responsible for the issuing of European Technical Approvals in the UK that lead to the CE mark. It also operates Approved Installer schemes.

Another conformance scheme is the Hallmark scheme, established for roofing personnel, and based on National Vocational Qualifications (Cook 1997). Loss Prevention Council (LPC)[6] certification is operated by the LPC, which is owned by the Building Research Establishment (BRE). The scheme certifies fire and security products and services, and is recognised in Australia by SSL.

Ecolabelling schemes include the Soil Association Woodmark scheme and the SGS Forestry QUALIFOR Programme, both accredited by the Forest Stewardship Council (FSC)[7].

In contrast to the BSI, the American National Standards Institute (ANSI)[8] does not develop standards. Rather it coordinates the activities of standards-setting agencies, and approves standards as American National Standards (Standards Australia now has a similar role). These are often joint publications, such as ANSI/AAMA 302.9 and ANSI/ASME A17.1 (AAMA[9] is the American Architectural Manufacturers Association and ASME, the American Society of Mechanical Engineers).

5. *See www.bbacerts.co.uk.*
6. *See www.bre.co.uk/lpc.*
7. *See www.fscoax.org.*
8. *See www.ansi.org.*
9. *See www.aamanet.org.*

ANSI represents American interests on ISO committees. The US has the secretariat for around 20 of 182 ISO technical committees overall; for construction-related committees, the figure is 5 of 35 (BSI has 5, SAA has 1 for solar energy). To give an idea of the scope of ANSI standards, in 1999 there were 24 listed under "door" on the ANSI website (Standards Australia's 1999 catalogue listed 14 under "Doors and windows").

The most prolific standards agency is the American Society for Testing and Materials (ASTM)[10], which had 32 titles under "door" on the 1999 website. A three-volume set of the 600 cited in building codes, among others, is available (*ASTM Standards in Building Codes*). ASTM is the source of most of the American standards cited by NATSPEC.

Other major agencies include ASHRAE (American Society of Heating, Refrigerating and Air-conditioning Engineers), NFPA (National Fire Protection Association), and Underwriters Laboratory (UL)[11]. The National Institute of Building Sciences (NIBS) publishes a set of CDs called the Construction Criteria Base, which is equivalent to a million pages of text. Information includes reference standards and the model codes. Standards bodies represented on the CD include ASTM, American Concrete Institute, Sheet Metal and Airconditioning Contractors National Association (SMACNA), American Institute of Steel Construction, National Association of Architectural Metal Manufacturers and Western Wood Products Association – about 115 organisations all up.

NIST, the former National Bureau of Standards, also publishes product standards, but this is not its main activity. With so many agencies producing standards, there is often more than one standard on a given topic e.g. ASTM, UL, NFPA and UBC, each with a standard test method for assessing surface burning. This means that the various codes will not necessarily cite the same standards for the same subject. The five standards most widely cited in the codes relate to structural steel, reinforced concrete, electrical work, surface burning characteristics and elevator/escalator safety.

On conformance, the federal government funds the Building Product Pre-Approval Program (BPPAP). This is run by NIBS and certifies conformance to federal specification performance requirements. In the future, an independent evaluation and listing program for building products and methods of construction will operate alongside the ICC. This program, to be called the National Evaluation Service (*NES*), will certify conformance to the *International Building Code*.

A number of standards-producing agencies, often industry-based, operate their own certification schemes. Indeed, the AAMA has offered an ANSI-accredited third-party certification scheme, for windows and doors, since 1972. Schemes are also run by (among others) the Certified Ballast Manufacturers (for fluorescent lamp ballasts), the Air Movement and Control Association (for fans and blowers), and Builders Hardware Manufacturers Association (for locks, latches, closers etc.).

Laboratory-oriented certifiers include Intertek Testing Services (which encompasses

10. See www.astm.org.

11. See www.ul.com.

ETL and Warnock Hersey), and NSF International, which certifies drinking water and water treatment systems and plastic piping components (Miller 1998). UL provides both a "listing mark" for product safety to UL standards, and a "classification mark" for performance to other standards e.g. ASTMs. So a product could carry two UL marks. UL listing is recognised in Australia by SSL.

Ecolabelling schemes include those accredited by the FSC: the Rainforest Alliance's Smart Wood Program, and Scientific Certification Systems (SSC) Forest Conservation Program. (SSC environmental product certification covers a wide range of products).

Master specifications

There are two complementary national master specification systems in the UK: the Royal Institute of British Architects (RIBA)'s National Building Specification (*NBS*), which covers built fabric, and Barbour Index/CIBSE's (Chartered Institute of Building Services Engineers) National Engineering Specification (*NES*), which covers building services and transport systems. *NBS* is published by NBS Services[12], a division of RIBA Companies.

NBS provides general advice at the beginning of each worksection, and detailed advice alongside the clauses, including cross-references to other sources. It is provided on a CD in database form as *Specification Manager*, and this includes management and editorial software, and hyperlinks to commentary, the full text of the Building Regulations and standards. It is also available in wordprocessing form (like NATSPEC) as *Specification Writer*, and this includes editorial software.

Codes and standards	Australia	United Kingdom	United States
Number of standards and codes of practice cited in building codes	209 (BCA including state and territory variations)	102 (Building Regulations 1985)	1700-2000 overall (survey of 53 codes); typically 300-400 per code
Number of agencies producing standards and codes of practice cited in building codes	13 (including Standards Australia, individual state and territory governments, CSIRO, ASTM and ISO)	1 (BSI)	91 overall (e.g. ASTM – 600 standards; ANSI – 30) (survey of 53 codes)
Total number of standards and codes of practice	6000+ (Standards Australia)	10,000+ (BSI)	9000+ (ASTM) 8000+ (ANSI) 70 (ASHRAE) 260 (NFPA) etc. 20,000+ overall
Number of building-related standards and codes of practice	1800 (Standards Australia)	1500 (BSI Handbook)	3000 overall 1370 (*ASTM Standards in Building Codes* Package)

Table 3.1: International codes and standards comparison

12. *See www.nbsservices.org.uk.*

Both *NBS* and *NES* are in one-part format. They provide prompts for project-specific text, and alternative clauses, requiring deselection of those that do not apply. *NBS* is available in standard, intermediate and minor works forms, and a landscape package is available.

Both systems use the Common Arrangement of Work Sections (CAWS) classification system for their material (see also Classification systems below). *NBS* does not have a standardised structure within worksections. *NES* worksections have four subsections, as follows:

- System objectives.
- Cross-references to worksections for common parts e.g. pipelines (Y sections in CAWS).
- Products/materials.
- Workmanship.

The main US private sector master specifications are Construction Sciences Research Foundation[13] (CSRF)'s *SPECTEXT98* and the American Institute of Architects' (AIA) developed *MASTERSPEC*. Both cover fabric and services and transport systems. *MASTERSPEC* is published by ARCOM.[14]

Neither of the main master specifications provides clause-by-clause commentary, though *MASTERSPEC* has an "evaluations" sheet with each section, covering standards, manufacturers, considerations, comparisons and technologies. Both provide editorial instructions embedded in the material. *MASTERSPEC* also provides a cover sheet (content, application and history), and checklists for coordination of specifications and drawings. Software is also available to simplify production, editing and metrication (SpecwareLinx from ARCOM, and Comspec from CSRF) (Bushnell 1999).

Both are only provided in one-part format. *SPECTEXT98* also provides options for selection and industry default language, and product choices have been added. *MASTERSPEC* is a genuine library of clauses and so relies mostly on deletion of unwanted alternative clauses.

Because they are both national systems, and because of the litigious nature of the US construction industry, they tend to cite all relevant standards. Specifiers must select, in part, according to the standards cited in their local code, and according to the standards used by local manufacturers. Therefore, unlike NATSPEC, *NBS* and *NES*, the master specifications cannot be tailored to a particular building code, nor (often) to particular standards. This obliges them to be fairly general in their coverage, and to require project specifiers to sort out any conflicts that might arise. National manufacturers deal with this by working to the most stringent codes or parts of codes. Specifiers cope by generally working only in one jurisdiction.

Both use the MasterFormat and SectionFormat classification systems for their material (see below).

13. *See www.csrf.org.*

14. *See www.arcomnet.com.*

Classification systems

The RIBA's Uniclass system of classification (Crawford et al 1997) is now being used across the UK. It is progressively replacing other systems.

Development of the Uniclass system was sponsored by four industry organisations (Construction Confederation, RIBA, Royal Institute of Chartered Surveyors [RICS], CIBSE), and the government (Department of the Environment). At least four existing schemes were used in its development: CI/SfB (Construction Index/Samarbetskommittén för Byggnadsfrågor), CAWS, CESMM3 (Civil Engineering Standard Method of Measurement), and EPIC (Electronic Product Information Co-operation).

The Uniclass system was also informed by work being done by ISO TC59/SC13 Working Group 2, which produced a framework for classification of information about construction works as one of a series of proposed international standards on the subject. The RIBA *Plan of Work*, the CIB (International Council for Research and Innovation in Building and Construction) *Master List*, and the *London Classification of Business Studies* (Aslib, The Association of Information Management) were also referenced. In short, Uniclass is neither a unilateral nor an arbitrary system of classification. In many respects it is an international system. The Uniclass tables include the following:

- Facilities.
- Construction entities.
- Spaces.
- Elements for buildings.
- Elements for civil engineering works.
- Worksections for buildings (CAWS).
- Worksections for civil engineering works (CESMM3).
- Construction products (EPIC).
- Construction aids.
- Properties and characteristics.

The main CAWS headings are as follows:

- A Preliminaries/General conditions.
- B Complete buildings/structures/units.
- C Existing site/buildings/services.
- D Groundwork.
- E In situ concrete/Large precast concrete.
- F Masonry.
- G Structural/Carcassing metal/timber.
- H Cladding/Covering.
- J Waterproofing.
- K Linings/Sheathing/Dry partitioning.

- L Windows/Doors/Stairs.
- M Surface finishes.
- N Furniture/Equipment.
- P Building fabric sundries.
- Q Paving/Planting/Fencing/Site furniture.
- R Disposal systems.
- S Piped supply systems.
- T Mechanical heating/Cooling/Refrigeration systems.
- U Ventilation/Air conditioning systems.
- V Electrical supply/power/lighting systems.
- W Communications/Security/Control systems.
- X Transport systems.
- Y Services reference specification.
- Z Building fabric reference specification.

CAWS uses a three-tier system for worksections, thus:

- H Cladding/Covering.
- H6 Slate/Tile cladding/covering.
- H62 Natural slating.

In the US, the Construction Specifications Institute (CSI)[15] develops (with Construction Specifications Canada), maintains and publishes MasterFormat (CSI 1995), which is used across North America for organising specification material (project information, products and work). It has 16 Divisions and a five-digit numerical code for worksections (e.g. 09220, for Portland cement plaster). CSI also publishes:

- SectionFormat (CSI 1997), a three-part structure for worksections which includes a standard list of articles i.e. clauses;
- PageFormat (CSI 1999), which provides a recommended arrangement of text on a page; and
- UniFormat (CSI 1998), which is used to classify elements and systems. It extends ASTM E1557 (1996).

The 16 MasterFormat Divisions for worksections are as follows:

- 1 General requirements.
- 2 Site.
- 3 Concrete.
- 4 Masonry.
- 5 Metals.
- 6 Wood and plastics.
- 7 Thermal and moisture protection.
- 8 Doors and windows.

15. See www.csinet.org.

- 9 Finishes.
- 10 Specialties.
- 11 Equipment.
- 12 Furnishings.
- 13 Special construction.
- 14 Conveying systems.
- 15 Mechanical.
- 16 Electrical.

SectionFormat's three subsections are as follows:

- Part 1 General.
- Part 2 Products.
- Part 3 Execution.

The *NATSPEC BASIC* worksections are listed in Chapter 4. Readers can do their own comparison between the UK, US and Australian approaches. Needless to say, other countries have yet other systems.

Metrication

Metric units became legal for scientific purposes in the UK in 1864 and for trade in 1897. There was then a gap in the process until a 1950 report recommended metrication (i.e. abandoning imperial units, rather than allowing dual units). The report, however, argued that the UK should wait until its major trading partners – then the Commonwealth and the US – had metricated.

A 1963 BSI survey found that a significant chunk of industry supported metrication. The Metrication Board, set up in 1969, was intended to see the UK largely metric by 1975, but this turned out to be realistic only for certain sectors of the economy. The Conservative government, unenthusiastic about metrication, scrapped the Board in 1980, but by this time many sectors, including construction, had gone metric.

There were EU Directives on weights and measures, in 1989 and 1991, and from 1996, all packaged goods in EU countries have been sold in metric quantities. From the end of 2009, all products sold in Europe will be labelled only in metric units, which means that the placing of Imperial 'supplementary indicators' after the metric quantity will no longer be allowed. This is a 20-year extension on the original Directive deadline, the extra time mostly due to trading with the US.

Curiously there is no obligation on the UK to ever change its road signs to metric units, though several factors (e.g. tourism) may eventually force it to go the whole hog, and metricate these too (Keenan 1999).

Metrication in the USA[16] has been even longer coming and, because the 50 states and different sectors of the economy all tend to do their own thing, is incredibly piecemeal. For example, the National Council of Architectural Registration Boards (NCARB) plans to

16. The US Metric Association website address is lamar.colostate.edu/~hillger.

have the Architect Registration Examination fully metric by 2004, but other players have set their own deadlines.

The principal piece of federal legislation is the *Metric Conversion Act* 1975, amended by the *Omnibus Trade and Competitiveness Act* 1988. This made the metric system the preferred system of measurement for US trade and commerce, and required that all federal agencies use the metric system in procurement and so on, as far as possible, by mid-1992. *Executive Order 12770 – Metric Usage in Federal Government Programs* encouraged government agencies to voluntarily adopt metric usage in procurement. But there has not been much progress to date.

Indeed progress occasionally goes backward, as in the *Transportation Equity Act for the 21st Century* 1998, which cancelled the Department of Transportation (DOT) year 2000 deadline for all states to use metric units in transportation areas such as federal roads and bridge construction, maintenance, and repair. However, metric is still permitted (and encouraged under the earlier Acts), and at least 21 state transportation departments have progressed too far to stop, and plan to continue converting to metric. This will have the effect of metric (in the transportation sector) states abutting inch/pound states, creating all sorts of problems.[17]

There is an international dimension to this inconsistency too. While the EU Directive mentioned above permits use of Imperial units as well as metric during the current transition period, these are often not the same as US inch/pound units, so this has not helped US companies very much. Conversely the 1996 amendments to the *Fair Packaging and Labeling Act* mandate dual inch/pound and metric units on goods sold in the US (imposing no time limit), putting European manufacturers at a disadvantage. With Europe mandating metric-only and the US mandating dual labelling, it could be argued that both positions constitute technical barriers to trade, so these current positions are likely to be modified.

One outcome of all this is that dual units appear in US codes and standards. Some are published in two versions: metric and inch/pound. Mostly, hard conversions are used – there has been little or no attempt to rationalise product sizes yet. Australian specifiers using US codes and standards need to be aware of this, and need to be familiar with conversion, especially because US units are not always the same as the Imperial units that were in use in Australia in its pre-metric days.

17. Perhaps the worst example of a problem arising from use of dual mensuration systems is the loss of the Mars Climate Orbiter, blamed on calculations for thrust done in pounds-force but fed into the Orbiter, without conversion, as newtons (Hawkes 1999).

18. See www.bre.co.uk.

Environmental building assessment schemes

The UK's Building Research Establishment[18] Environmental Assessment Method was developed in 1990. There are five different schemes relating to new office designs, housing designs (*BR389: EcoHomes - the environmental rating for homes 2000*), superstore designs, industrial unit designs and existing office buildings. BREEAM's aim is to set criteria for good (defined by BREEAM) environmental performance in buildings.

The assessment is carried out at the design stage or, for existing offices, during a building survey. BREEAM is an example of a pragmatic approach to uncertainty on environmental and health issues – it steers clear of topics that its authors believe are unresolved. About 30% of new offices are assessed using the Method (Doggart 1996).

The design or building is assessed against a set of defined criteria and credits are awarded where these criteria have been met. Issues are divided into three categories: global and use of resources, local and indoor. They include the following:

- Carbon dioxide emissions resulting from energy use in the building.
- CFC (chlorofluorocarbon) and HCFC (hydrochlorofluorocarbon) emissions.
- Use of natural resources.
- Renewable, non-renewable and recycled materials.
- Water economy.
- Ecological value of the site.
- Cyclists' facilities.
- Hazardous materials.
- Thermal insulation.
- Environmental and purchasing policies.

An overall rating of Fair, Good, Very Good and Excellent is awarded, depending on the number of credits achieved under global, local and indoor categories.

The US Leadership in Energy and Environmental Design (LEED)[19] is a whole building assessment scheme managed by the US Green Building Council. It is designed for new and existing commercial office buildings and high-rise residential. Other buildings will be covered under a separate rating system. Green building certification is awarded to buildings on a scale of platinum (highest) to bronze (lowest), depending on the number of credits achieved. Credits are assessed against set criteria.

Prerequisites for rating include asbestos avoidance, indoor air quality, lead in water, and erosion control. Credits cover topics including building commissioning, indoor air quality, building materials (VOC content, recycled content, resource use), construction waste management, energy, use of existing buildings, energy efficiency, landscaping, recycling facilities, operations and maintenance facilities, ozone depletion, siting, transportation, water use and water quality.

Project process

The process and the scope of the architect's services are much the same in the UK and the US as in Australia (see Chapter 2). In the UK the process is usually presented using the RIBA's *Outline Plan of Work 1998*, which can be adapted for different forms of procurement. The main stages of the *Plan of Work* are as follows:

- Appraisal.

19. See www.usgbc.org.

- Strategic brief.
- Outline proposals.
- Detailed proposals.
- Final proposals.
- Production information.
- Tender documentation.
- Tender action.
- Mobilisation.
- Construction to practical completion.
- After practical completion.

Element	United Kingdom	United States
Tendering documentation	NJCC (National Joint Consultative Committee for Building) Codes of Procedure e.g. *Single Stage Selective Tendering*, 1994. Supplementary material from *NBS* and *NES*	*SPECTEXT98* Series 0 – Bidding and contracting requirements
General conditions of contract	e.g. JCT contracts PWQ98, and WCD98, ICE (Institution of Civil Engineers) *The New Engineering Contract* 1993	e.g. AIA A201 (1997), EJCDC (1997)
Special conditions of contract	Presumably as Appendices to JCT forms – no templates provided. Selection among main and secondary options for ICE	Supplementary Conditions – guidance only, no templates e.g. AIA A511, and *Guide to the Preparation of Supplementary Conditions*, respectively
Specifications – Preliminaries Specification – General requirements	Material from *NBS* and *NES Preliminaries/General conditions* Material from *NBS* and *NES*	*MASTERSPEC* and *SPECTEXT98* – Division 1 General requirements
Specifications – Trades or technical material, also known as worksections Specifications – Multi-trade schedules	Material from *NBS* and *NES* No source	*MASTERSPEC* and *SPECTEXT98* – Divisions 2 to 16
Drawings	BS 1192 series	CSI *Uniform Drawing System.* AIA *CAD Layer Guidelines*
Bills of quantities	RICS *Standard Method of Measurement* (1998)	Not used in the USA. CSI UniFormat used for classifying elements

Table 3.2: Contract documentation elements and their sources – United Kingdom and United States

A number of clauses relevant to specifiers in UK and US contracts are discussed in later chapters. The main point to make about contracts in regard to contract documentation is that conventional construct-only contracts such as the JCT's PWQ98 (1998), and AIA A201 (1997) tackle the issue of design by the contractor, so permitting use of performance specification. This will be discussed in more detail in Chapter 6.

Vocation and training

The longstanding British text, *Specification Writing for Architects and Surveyors* by the quantity surveyors Willis and Willis, is in a new (1997) edition, which was roundly criticised in *The Architects' Journal* (Whiteley 1998). An alternative, *CPI: Project Specification – A code of procedure for building works* by the Association of Consulting Engineers (ACE et al 1987), is a bit long in the tooth, but as part of a set is an interesting guide to recommended UK documentation practice generally. It is certainly more relevant than Willis and Willis, which offers relatively little general advice, and is being updated.

Specification writing does not seem to be taught to architects to much of an extent at undergraduate level in the UK (McGregor 1999). This may reflect the tradition of QS-written specifications.

The most significant specification resources in the US are those published by CSI for its over 17,500 members. Its *Manual of Practice* (CSI 1996a and 1996b) comprises four modules:

- Fundamentals and formats.
- Specifications practice.
- Construction product representation.
- Construction contract administration.

CSI uses the *Manual* as the basis for specifier accreditation and training and for judging its annual specification competition. CSI also publishes the monthly magazine, *The Construction Specifier*. This covers context, building products and systems, specification design and planning, contractual, technical, and editorial matters, use of specifications, and the future of specifying. Other CSI publications include the *Uniform Drawing System* (UDS), *Spec-Data* (for product information), *Manu-Spec* (for proprietary specifications)[20], and convention cassettes.

Rosen and Heineman's *Construction specifications writing* (1990) (updated by Rosen [1999]) is a recommended alternative to the CSI *Manual*. Both documents are quoted in RAIA literature.

CSI has an active student program but the impression from the literature is that specification writing is not taught to any great degree at undergraduate level.

20. The UK equivalent is NBS Plus.

Summary

Broadly speaking, specification writing is about as complex in the UK as in Australia (e.g. regional variations to the building code), and much more complex in the US (e.g. many standards-producing agencies and building codes). But in both countries the documentary resources available to specifiers are far more extensive than in Australia. The master specification systems are more comprehensive, and there are many more standards. The Internet is much more widely used, particularly in the US, for product information, legislation, proprietary specifications and so on.[21] Also, the construct-only contracts tackle contractor-design effectively, whereas those in Australia do not.

Manufacturers seem to take the standards more seriously (e.g. citing them in their advertisements). They also take the specifiers more seriously. The markets of course are bigger and the range of products and services on the market is far more extensive than in Australia, so there is more competition.

In the UK, Europe is becoming more influential through various EU Directives. Conversely, the UK has considerable influence in Europe, through BSI for example. In the US several strands of rationalisation are in hand – notably metrication and a single national building code – which should make the US construction industry more efficient and clear the way for it to become an even bigger player internationally e.g. in development of standards.

In the US (and Canada), CSI is very influential. Thanks to the CSI membership policy of encouraging product sales representatives to join as well as specifiers, manufacturers take specifiers' needs more seriously. Websites are full of proprietary specifications in standard format. CSI's engineering specifier membership means that services specifications are generally better in the US than in Australia or the UK. Indeed the existence of CSI, which has a handful of Australian members, must be beneficial to specification production all round, through its magazine, conventions and competitions, for example. Australia would do well to consider establishing a CSI Chapter.

21. Ranking by country of web addresses registered in 1998 is as follows:

1. *US.*
2. *Canada.*
3. *UK.*
4. *Germany.*
5. *France.*
6. *Sweden.*
7. *Spain.*
8. *Italy.*
9. *Netherlands.*
10. *China.*
11. *Japan.*
12. *Australia.*

About 70% of new registrations were for businesses and individuals in the US (IHT 1999).

4 Programming and specification design

As suggested in the Preface, the production of a specification is a project in its own right, embedded in other projects like a Russian doll. Like all projects it will benefit by time spent in planning before it gets underway. This chapter looks first at the planning of the process of specification production, including sources (of information), resources including time and personnel, tasks and their sequence, and consequences of applying quality assurance to the process.

It then looks at the design of the specification – the product – itself. Here we concentrate on the general issues of thoroughness and structure. Following the documentation principle that you should say it once, and in the right place, the next thing to resolve is what will be on the drawings or in the contract, and what will be specified. This will be discussed in the next chapter.

Thoroughness, or the quality of the content of the specification itself, will vary from project to project. This chapter outlines points to be considered in deciding the appropriate level of quality. The structure of the specification is discussed at two levels: overall, and within each worksection. Overall structure may vary somewhat from project to project, especially in terms of worksections included or excluded, but should follow the general principles outlined in this chapter. Structure within worksections should be dealt with consistently from worksection to worksection. Several alternative approaches are given.

Sources

Specifiers can choose from five sources for a new project specification. They can

- use raw data;
- adapt a previous project specification;
- use a proprietary specification;
- use an office master specification; or
- use a commercial national master specification system.

No matter which approach is used, culling unwanted material and being technically up to date are the main issues, especially with pre-edited material. Also important is consistency with other specifications emanating from the practice.

Using raw data will often be necessary for the odd clause or section where new or unusual (at least, to the specifier) material or component, and associated workmanship, needs to be described. The research function of the specifier is underrated. But it is not necessary to go to raw data for standard materials or components because that would involve reinventing the wheel – and the doors, the paint, and the concrete.

Specifiers needing to start from scratch are advised to investigate thoroughly. Are there any standards (not necessarily Australian) that could help? Is there an industry organisation that can offer impartial technical advice? Are there any up-to-date journals or texts on the subject (as there are for straw bale construction and photovoltaics), for example? What trade literature is available?

Are there any specifications for the subject on the Web? Some websites carry downloadable proprietary worksections that can provide a good starting point for a generic

Specification source	Advantages	Disadvantages
Raw data	• Unavoidable sometimes	• A lot of work
Previous project specification	• Quick	• Currency • Relevance • Inherent faults
Office master specification	• More likely to be current • More likely to be culled/supplemented for relevance • Faults likely to be fixed centrally • 'Expert' specifier in charge • Embodied office knowledge	• Time-consuming maintenance • Can only draw on office expertise • Must be learnt by others • Not consistent with co-consultant masters
National master specification system	• Low development and maintenance costs • Draw on a wide range of resources • National 'expert' specifiers in charge • National best practice • Covers all worksections for consistence right across the specification • Clients, contractors and other players already familar with it	• Must be learnt • Will not reflect office, client or project type issues • Office knowledge must be added

Table 4.1: Advantages and disadvantages of various specification sources

worksection.[1] Are there other specifiers who may have dealt with the problem before (computer bulletin boards may help here[2])? Wherever possible, manufacturers, fabricators and contractors with relevant experience should also be consulted.

The NATSPEC *Earth block walling* worksection, for example, used several US prescriptive building codes, taken off the Web, and a US and an Australian guide to specifying the material, to get to first draft. There are no proprietary specifications for this work. There are New Zealand standards but the opportunity to adopt them as joint AS/NZ standards was rejected, which implies that they have limited applicability in Australia. The NATSPEC *Fire-stopping* and *Glass blockwork* worksections used several proprietary specifications taken off the Web, and in hard copy, from Australia and the USA. *Fire-stopping* also relied on an Australian standard, AS 4072.1 (1992), and a BCA deemed-to-satisfy specification (C3.15). Worksections in other master specification systems (from the UK and the USA) were also referred to for these two worksections.

Adapting previous project specifications – what is sometimes called producing 'rolling specifications' – is a popular, if hazardous, approach. It is popular because it saves a great deal of time, by minimising reinvention of wheels and the like. It is hazardous because of the risk that text irrelevant to the project is retained and text relevant to the project is not included. This might occur if the source specification was not checked to see if irrelevant text had been added and relevant text deleted, a common mistake given the hurried circumstances in which rolling specifications are used.

There is also the problem of updating the old specification material in terms of standards, and statutory and technical (e.g. brand names) changes. Indeed such specifications are a bit like the immortal soup of the folk tale – there will always be a morsel of the very first specification in the current version. A number of specifiers manage their project specifications with this use in mind, using a database system for filing clauses. Nevertheless, the fact that such specifications are turned around as quickly as possible militates against a thorough job. On balance, such an approach cannot be recommended (see Figure 4.1).

And yet, as you might expect, rolling specifications are still widely used. This approach tells us that little time (and care) has been spent on the document. It is not seen as worthy of much effort. It is as if the specifiers are asking, "who'll use it anyway?" and perhaps they are hoping, nobody.

Manufacturers are increasingly making available prototype specifications geared to use of their products. Several, such as those produced by Dulux, are very extensive comprising, in this case, a general specification and a series of product specifications. Proprietary specifications are a useful resource, but must be used critically, even if you want to specify the proprietary items covered. There are several reasons for this:

- They do not use a consistent format, in particular they are rarely aligned to NATSPEC (and any claim that they are should be taken with a grain of salt).

1. For example, Dulux at www.dulux.com.au, and from the USA, Construction Zone at www.c-z.com, Arcat at www.arcat.com, and 3M at www.mmm.com/market/construction/html.

2. Membership of the Construction Specifications Institute and/or Specifications Consultants in Independent Practice (SCIP) gives access to US self-help material over the web.

- They may omit material that would provide protection against deficiencies in the products – they are promotional, after all.
- They are often poorly written and poorly edited. For example, they inevitably address subjects that are either covered elsewhere or should not be addressed at all (e.g. contractual material, construction methods), and often speak to the subcontractor rather than the contractor.
- They may be silent on general matters such as inspections, tests and maintenance manuals.

Of course, if one wants to specify generically, then proprietary specifications may offer little assistance. For example (particularly in Australia), they rarely refer to the relevant standards or to the availability or otherwise of product certification.

Many offices work on the same type of project repeatedly, sometimes for the same principal e.g. bank fit outs, schools, Aboriginal hostels. A carefully prepared master specification, a logical development of the 'rolling system', maintained as a separate entity and tailored to the project type, is a valuable time and risk-saver.

Figure 4.1: The trouble with rolling specifications

The office master will, of course, contain a great deal of standard material e.g. masonry specified to AS 3700 (1998). But it will also contain default material particular to the project type or principal, such as the preferred lock system specified by proprietary item, or all reference to box gutters deleted (and a guide note to that effect). Only the master need be maintained, rather than a series of predecessors, and this is much easier than dealing thoroughly with a 'rolling system'.

Offices with such systems can find them difficult to maintain and resource. Even large public works offices under-resource their master specifications, sometimes to the extent of abandoning them altogether. Recent examples are the Victorian and Tasmanian public works authorities, which both switched to NATSPEC in the mid-1990s. This problem arises elsewhere. The Canadian government's master specification (*National Master Specification*) was staffed by one professional specifier in 1998, with a master of over 120 worksections to look after, just under Sitework!

Carrying the idea of an office master one step further, many countries have established commercial national master specifications (McGregor 1997).[3] The advantages of such masters are numerous, for they

- provide for uniformity in presentation and formulation of project specifications;
- allow standardisation of terminology, definitions and interpretations with the aim of these being widely understood;
- keep specifiers up to date with changes in legislation, regulation, standards, technology, products and working practice;
- achieve economies in specification production by eliminating the need for specifying organisations to create and maintain all the specification material they use;
- provide prompts, advice and information which may be of assistance in the preparation of specifications;
- reflect a national consensus about what comprises sound standards of materials and workmanship in contemporary building practice;
- encourage production of clear, concise, consistent, well-written and easily interpreted specification clauses;
- raise the principal's confidence that the specification and the building will be of a reasonable quality (a form of quality assurance); and
- minimise, through the consensus approach used in their preparation, application of the *contra proferentem*[4] principle since all sides of the industry have contributed to the material.

Of course, there are disadvantages. Since such masters are complex tools prepared by someone else, project specifiers have to become thoroughly familiar with the material. Like all tools, master specifications can be used or misused, and so usually come with a set of instructions, qualifiers, and so on. In particular, a certain minimum level of skill and interest

3. The International Construction Information Society (ICIS) is an organisation of national master specification publishers. See www.icis.org.

4. Contra proferentem is the ancient legal principle that ambiguities in documents are, generally, to be construed against their author. For custom-made construction contracts and specifications prepared by the principal, this would mean that the courts interpret ambiguities in favour of the contractor.

In this respect, the advantages of consensus documents are that

(a) the principal did not write them; and

(b) because the authors included representatives of principals in general, and of contractors in general (and maybe of contract administrators in general), the text is as fair as it can be.

Ambiguities are therefore sorted on a case-by-case basis.

has to be assumed. Skipping the learning process under pressure of time, by not reading the manuals or perusing the text and commentary, is counterproductive in the long-term.

The defaults of the national masters may not suit you, the options may not include your preference, and the coverage may be less than you want. The masters are likely to be generic, for example, so proprietary items will not be covered (though more national master specification systems are providing proprietary specifications e.g. New Zealand's CIL *MasterSpec*, UK's *NBS Plus*, US's *Manu-Spec*). They are likely to assume new-build, so fit out and restoration may not be covered. They are likely to cover only core work items, so unusual work such as photovoltaics may not be included.

Finally, Standen (1995) argues the case for the "human side of specifying", and seems to be against the use of national master specification systems.

It is possible to adapt national masters, converting them to office masters, through a process of pre-editing. But the maintenance provided by the master specification publisher will not cover pre-edited material, making maintenance of that material tricky.

The Queensland Department of Public Works (1996) and NSW public works authorities (DPWS 1997) have found it difficult to adequately resource their own NATSPEC-based masters e.g. for schools and correctional services buildings. Consultants might take on more of this work as has already happened in Queensland.

Resources

No Dispute (see also Chapter 2), includes the following recommendations:

- 3.5 "Adequate time and resources should be allocated to the design and documentation process".
- 3.9 "Quality Assurance and Control principles should be applied to the design and documentation process".
- 3.6 "Adequate services and fees should be agreed by the Principal for the consultancy services" – a necessary corollary to 3.5 and 3.9.

Clearly this general advice applies to specifications.

Specifications take time to write, even those based on tailored office masters. For smaller projects, say around $1 million, a thorough specification may take up to 15% of the total documentation effort. For large projects, say $100 million, this reduces perhaps to around 2.5%. Of course these figures are crude. Time taken depends on the source of the specification, thoroughness, expertise, the timeliness and adequacy of input from other consultants, not to mention the complexity and size of the project.

The Australian Institute of Quantity Surveyors (AIQS) 1995 fee scale explicitly listed quantity surveyor's fees for specification preparation. They assume that the architect provides thorough notes. The figures for weeks are derived from the percentage fees, assuming a $100 an hour charge-out rate and 37.5 hours per week:

• Tender net cost of works ($M)	1	5	10	50	100
Specification fee (percentage of contract sum)	1	0.35	0.20	0.06	0.05
Specification – weeks	2.7	4.7	5.3	8	13.3

For architects, the fee percentage for contract documentation may be 40% of the total project fee. Working from this using the fee chart for conventional works in RAIA Advisory Note AN02.03.100 (1996a) and assuming a sliding specification percentage from 15 to 2.5 as suggested above (so that, as has been noted in Chapter 2 , between 1% and 6% of architectural fees are earned through the preparation of specifications), the above table for specification preparation becomes something like:

• Cost of the works ($M)	1	5	10	50	100
Specification fee (percentage of contract sum)	0.44	0.28	0.19	0.10	0.04
Specification – weeks	1.2	3.7	5.2	13.3[5]	11.3

Programming the specification is obviously not an exact science – the two tables are not in perfect agreement (Figure 4.2). Assuming these figures (for quantity surveyors and architects) are not complete flights of a fee-setters fancy, to keep a full-time specifier busy (44 weeks productive time) an office would need eight $10 million projects, or at least 16 $1 million projects, all reaching the tendering phase. Specification writing is therefore almost invariably a part-time affair, as few offices are this busy.[6]

The point here is that specifications take time. Yet NATSPEC material is often requested only days before the specification is meant to be finished, along with the rest of the documentation. This doesn't leave much time to learn how to use the NATSPEC system, let alone resolve the technical issues and so on. Clearly many people leave the specification till as late as possible.

Who should write the specification? The specifier should have several qualifications:

- Familiarity with the project, including design goals.
- Expertise in contractual, technical and editorial matters (AACA *Competency Standards* may offer a few clues here).
- Willingness and availability.
- Cost-effectiveness i.e. minimum product of charge-out rate and time taken.

Options include non-specialists or specialists (including subcontract and consultant specifiers), architect or quantity surveyor (or engineer) as lead specifier, and status of specifier e.g. junior, director, or a person in between. For smaller projects the project architect is probably the best choice as lead specifier, but is unlikely to be available on larger projects. An internal person of similar status is generally the next best choice, ideally someone involved already in design and/or documentation of the project. Specialist specifiers are a good choice if available, but subcontract specialists (i.e. specialist specification writing consultants) cannot make design decisions without approval. They need thorough briefing (Heineman 1999a; 1999b).

5. The RAIA figure for $50M, showing that it takes longer to write a specification for a $50M project than a $100M project, is a result of the original data not the author's calculations.

6. A 1992-93 Australian Bureau of Statistics (www.abs.gov.au) survey (ABS 1992-93) of 4409 architectural practices (including town planners, architectural drafters, landscape architects) in Australia showed that 80% (up from 75% in 1988) comprised 4 staff or fewer. Only 1.8% had more than 20 staff (earning 29% of the income). A later survey of engineering practices (including quantity surveyors) shows that they follow a similar pattern (ABS 1995-96)

Standen (1995) draws attention to the law of some states which stipulates that only lawyers, or the parties concerned, can draft contracts – suggesting that writing specifications (and preparing working drawings?) should therefore be done by lawyers. This seems to overlook the point that architects and others who routinely prepare such documentation do so in part as agents of the principal, one of the parties concerned.

On the use of quantity surveyors it has been suggested, by a quantity surveyor,[7] that they may know more about the contractual issues than do architects, but conversely know less about the technical issues. Perhaps a quantity surveyor could advise on the preliminaries (about which, more later), for example, but generally the architect and structural engineer would certainly be best on the non-services technical sections. Whiteley (1998) suggests that quantity surveyors have limited vision; they "perceive the purpose of specification only in relation to construction (and tender) documentation". As discussed in the Preface, the specification has several other functions.

On cost-effectiveness, while senior staff would perhaps be able to do the job quickly, this is offset by the high charge-out rate. Their work should still be checked in spite of their status – it may have been a long time since they wrote a specification. Conversely, juniors with a low charge-out could be slow and unreliable, with comprehensive checking and substantial reworking almost certainly being required. Someone who can do the work fairly quickly, reliably, and at moderate charge-out, is required. This principle applies, of course, to all work

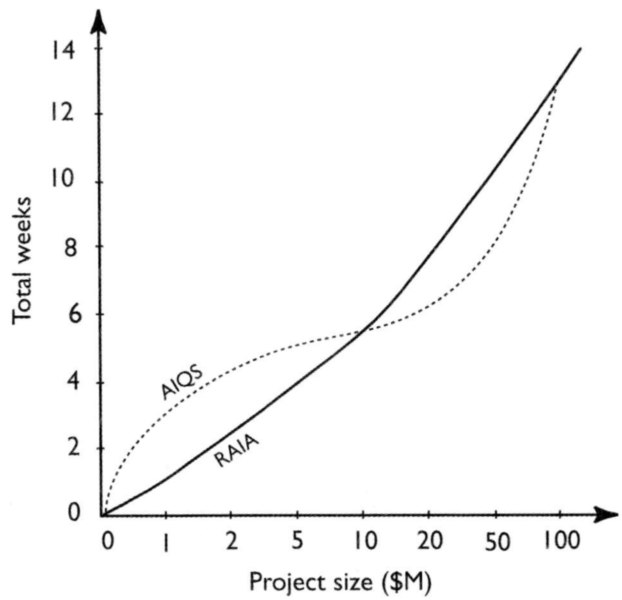

Figure 4.2: Fee-based time allocation for specification writing

7. Personal communication with Peter Clarke.

in the office, but the ideal person will vary depending on the task (e.g. dyeline copying versus client liaison). It will also depend on whether the project is time or cost-critical.

This person may not be available, however, and all too often, availability alone dictates selection of the specifier, particularly because time is short. Accordingly, specifiers are commonly inexperienced and/or unwilling or busy on other tasks. Also, the specifier is often neither the contract administrator nor the designer, creating problems with continuity. The specifier is sometimes an 'expert' but this is no guarantee of a good job. Indeed, this may only serve to perpetuate bad habits (some such experts are old dogs, and it is hard to teach them new tricks).

Whoever the project specifier is, it should be borne in mind that the description of the works is a statement of the principal's contractual requirements, not of the whim of the designers. The specifier, like the rest of the documentation team, is merely acting on the principal's behalf.

Program and process

Production of the specification should proceed steadily from preparation of the building approval (compliance) specification and, as already indicated, should not be postponed until a last minute panic. Specifiers should have a great deal of input into the documentation process. In particular, they can resolve what will be shown on the drawings (as drawings or notes) and what will be specified, coordinate terminology, and provide a checkpoint and prompt for detail design decision-making.

The specifier should not merely take instructions on this. The specification, if continually worked on, can also be used by estimators at the various check-estimate points e.g. 20%, 50% documentation. Other project personnel, including co-consultants, should be informed of the proposed content of the specification early in the process.

An aid to programming is the Pareto principle or 80/20 rule (McConnell 1989).[8] This suggests that there is a non-linear link between effort and output. In particular, 20% of the effort will produce 80% of the output. The remaining 20% of output will require 80% of the total effort (a case of diminishing returns)! The issue becomes one of knowing when to stop – it is rarely worth the effort to produce 'complete' documentation. This 'omitted' work needs to be built into the cost plan – it still has to be built, after all, and is perhaps covered in contracts at, for example, JCC clause 6.01 (as obviously inferred minor items of work). One should perhaps aim for 90% or so complete as a fair compromise of output and effort (Figure 4.3).

The following 13 general steps in the process of preparing a specification assume that a master is being used (see also Figure 4.4):

1. Update the source master copy, both in print and on disk. This should be done regularly, not just when you are about to start a new documentation project (if these are rare, the updating is likely to get out of hand).

8. This principle was named for Vilfredo Pareto who, while studying distribution of wealth in Europe in the late 18th and early 19th centuries, concluded that 80% of the wealth was in the hands of 20% of the population. Joseph Juran, a 20th century total quality control guru, showed that the principle has wide application in quality control e.g. 80% of mistakes are produced by 20% of causes. The Pareto principle certainly helps to focus one's attention (McConnell 1989).

2. Pre-edit the source for common project classes (e.g. hospitals) or principals (e.g. preliminaries material), if appropriate, creating a submaster in the process.
3. Reformat the source if necessary – different practice principals (or directors) may have different requirements, for example. This could be done towards the end of the process. Create a project copy.
4. Decide on the working medium i.e. will you work direct on-screen or by marking up photocopies? It may be that the initial run-throughs, which generally involve deleting large chunks of unwanted material from the source, can be done most efficiently on-screen, and that more detailed work can be done best on printed copies of this briefer working document.
5. Decide on the working pattern. Will you run across all worksections one step at a time, keeping them all in parallel states of development (this is very difficult to do)? Or will you concentrate on one worksection at a time, or come up with a combination? The pattern will depend in part on available information and on the documentation program. For example, you may not make final decisions on cladding until late in the

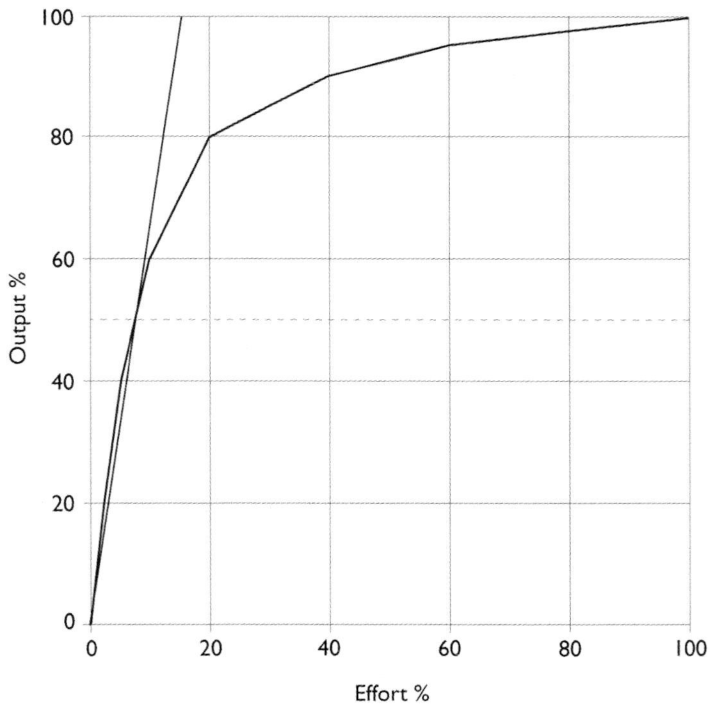

Figure 4.3: The Pareto principle, or the 80/20 rule. Programmers often assume a linear relationship between effort and output, and indeed it holds true for the first 50-60% of output. But after that point output rapidly diverges from the linear relationship, making a mockery of any extrapolation from progress at 50% of output.

process, but are able to almost finish the partition specification early on. Alternatively you may be obliged to complete a package, such as windows (due to lead times off-site), early in the process, and yet postpone the start of the painting package.
Try to be systematic, so you don't get lost. Keep records of what you have done and plan to do. This is essential, due to the often intermittent nature of the task. Program the specification work with respect to the drawings program (the metalwork specification cannot really be sorted until the balustrades are designed and drawn up, for example). Also consider time and resources, such as the budget, number and grade of available personnel, and access to computers. Ideally, work on the specification in one solid effort, rather than stopping and starting because of other demands on your time.

6. Identify sections in the source specification that are not required, and delete them. This may drastically reduce the size of the working document, particularly important if you are working on a paper copy. Leave until later, and flag, those sections about which you are uncertain. This is the first step in the production of a project-specific document. Similarly, identify and delete subsections and clauses that do not apply. From now on it is probably best to concentrate on particular sections, rather than jumping from section to section. Print the edited document if you have not already done so, because it will probably be easier to work on the printed version from now on, rather than on-screen. Identify and delete subclauses and other small bits of text that do not apply.
7. Now begin adding material, though this could be done in parallel with the previous step. If the source contains prompts for project variables such as brick type or paint colour, complete these prompts next. Bear in mind that not all prompts will be relevant to the particular project. Check against other technical literature, particularly for product availability. Begin with the sections about which you know most. Use tables (i.e. schedules) where appropriate – these can be very useful for presenting complex information.
8. Add new material to the draft specification where required – this will be material not anticipated by the source. Check for variables that need to be specified e.g. alloys, thickness, colour and finish. This material may be added as a new worksection, subsection, clause or whatever. Locate it in a rational manner, matching the general sequence of the source i.e. do not put it all at the end!
9. Edit the remaining source material systematically – part of it is bound not to apply as it stands to the project in hand. There may be conflicts with new text or the drawings e.g. specification of a proprietary item will mean that corresponding generic material in the source may become irrelevant. Check thoroughly that items drawn are covered by the specification somewhere, and vice versa.
10. Repeat the above project-specific steps as often as required until the specification is finished. Ensure there are no items about which you are uncertain – if in doubt at this stage, leave them out. These items should have been flagged in earlier steps.

11. Check cross-references, both internal and external e.g. to standards and drawings (particularly door and window codes). Eliminate unnecessary repetition between the specification and drawings, and within the specification. Provide a summary list of standards actually referenced by number in the specification, to assist users of the specification and as a check against errors in citation. Such a list is for information and so should not form part of the contractual specification.
12. Renumber the final document – the source's original numbering may no longer apply. Use consecutive numbers, for navigation but not for cross-references.
13. Finally, print and proofread, perhaps in several iterations. Check the format, number the pages, finalise the contents list, prepare an index, and print and bind the appropriate number of copies. Distribute as required e.g. to principal, consultants, tenderers. Retain working and library copies.

The reality is that many specifications are prepared in such a hurry that a systematic approach, such as that outlined here, is out of the question. This may well change as offices move towards quality assurance certification.

Quality assurance

Quality assurance requires the documentation of critical procedures, and adherence to these procedures. It also suggests commitment to continual improvement, and to maintenance of tools and resources. Thus quality assurance considerations lead to the training of specifiers and the updating of specification source material.

Each specifying practice sets up its own procedures. There are no hard and fast rules, but there are texts that can be referred to for guidance. *Competency Standards* (see Chapter 2) are not in themselves procedures. Charles Nelson (1988, out of print) briefly addressed the process of specification preparation and NATSPEC publishes recommended procedures and worked examples for the use of subscribers.

The RAIA Advisory Notes AN14.03.100 (1994b) and AN04.02.101 (1990a) offer rudimentary advice on procedures, as does the RIBA *Architect's Job Book* (Cox and Hamilton 1995) at *F-G/S2: Production information: Specification and Schedules of work.* The CSI *Manual of Practice*, particularly the specifications practice module (1996b), provides a thorough set of procedures (see also Chapter 3). For generic advice to specifiers, see also Kalms (1990).

Updating a specification includes incorporating analyses of specifications prepared for previous projects (i.e. feedback, such as how many variations could be traced to the specification?) as well as tracking standards, regulations and changes in technology and practice. Australian and other standards can be tracked directly through the monthly magazine *The Australian Standard*, published by Standards Australia.

Standards Australia also offers services that automatically maintain and update a customer's library of Australian and international standards. Amendments to Australian standards can be downloaded free from their website. Tracking overseas standards can also

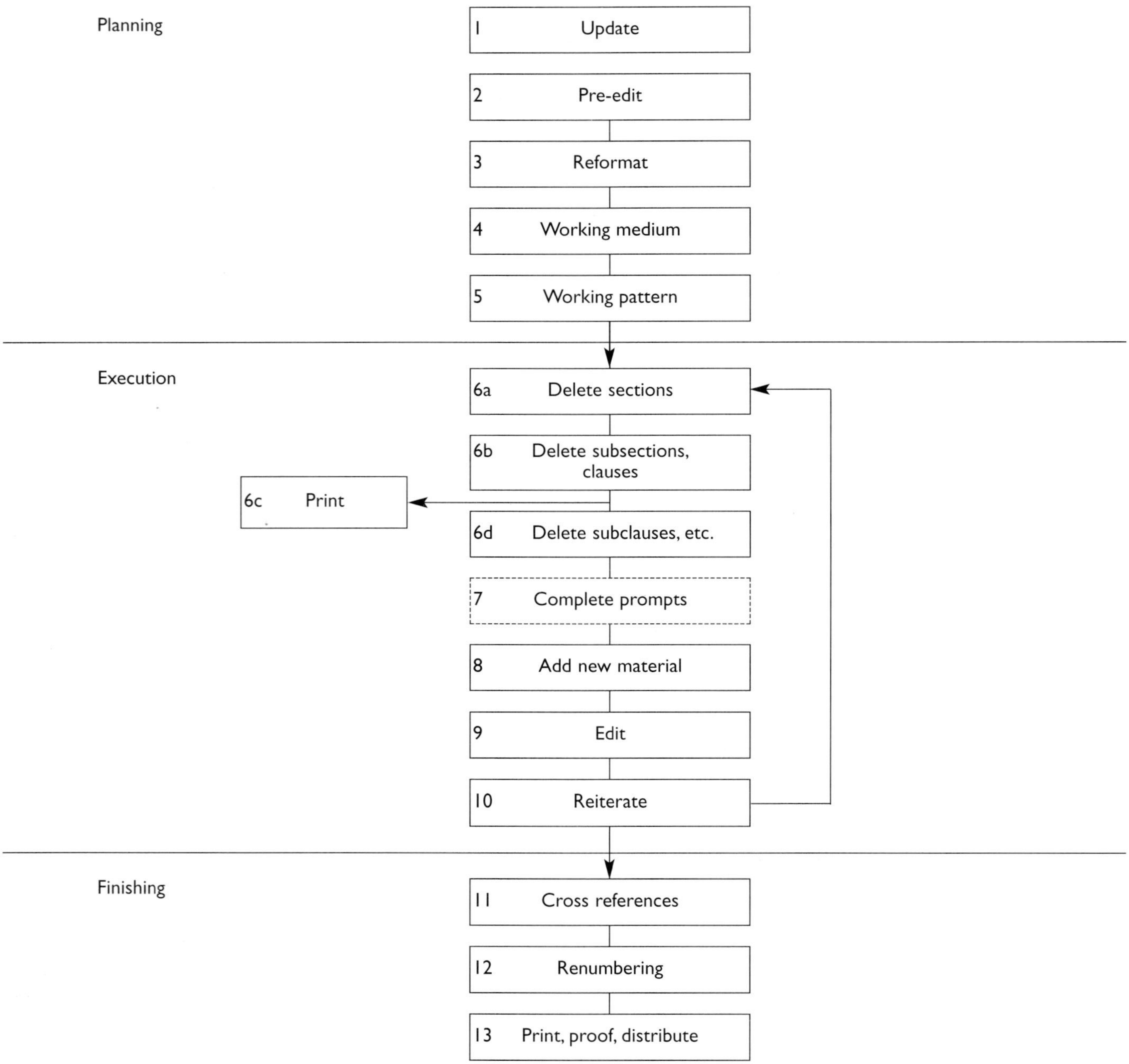

Figure 4.4: Thirteen steps in specification writing. Apart from step 7, these steps apply to the recycling of previous project specifications, as well as the use of office or national master specifications. Step 7 only applies to the use of master specifications.

be done through their publishers' websites (see Chapter 3).

Changes in regulations and the BCA can be followed through the various subscription services available. NATSPEC subscribers receive six-monthly updates, which pick up most relevant changes. The RAIA provides an updating service to Practice Services subscribers, for the Advisory Notes.

Updating is evidently a low priority for a number of specifiers – it too is mainly done at the last minute, if at all. So, the specification cannot be well-meshed with other documentation and the work of other disciplines.

For all specifiers, keeping track of standards, regulations and technology is a major problem, particularly in response to shifts in emphasis brought about by society's concerns e.g. embodied energy, occupational health and safety. It can take even longer to get used to radical shifts, such as adoption of performance-based codes.

Thoroughness

The quality of the specification's content is really its thoroughness[9] and is distinct from the quality level of the items specified (dealt with in Chapter 6). A thorough specification (i.e. where requirements are spelt out explicitly, rather than relying on the implicit or unspoken) could set low quality levels, and vice versa. The thoroughness of the specification depends on the following factors:

- Time available for documentation.
- Thoroughness of the drawings.
- Use of innovative versus conventional construction.
- Use of quality assurance versus quality control on-site.
- Whether or not the designers and/or documentors will be following through on-site.
- Level of trust or confidence in the contractor.
- Level of risk prepared to be assumed by the principal.
- Project quality level.
- Project cost or size.
- Fussiness of principal.
- Level of annotation on the drawings.

Clearly the less time there is to prepare the specification the less thorough it can be. To avoid the all-too common position of having time dictate quality, start the specification early and continue working on it steadily through the documentation program. Do not leave it till last. On the other hand, do not resolve worksections too soon. You may have to undo much of what you have done, as detailing progresses. Specification progress should parallel that of the drawings, though unfortunately some specifications are prepared in such a hurry that blanket reference to master specification or reference material, for example, is made without any qualifications – a very undesirable state of affairs.

It could be argued that the thoroughness of the specification should match that of the

9. What is thorough for a garden shed (where fixings can be specified simply as 'galvanized' if at all) is not going to be thorough enough for a nuclear power station (where fixings may be described in terms of strength, durability, compatibility and so on).

drawings. It is unlikely that shoddy drawings can be rescued by a very thorough specification, and vice versa – generally they convey quite different sorts of information. The specification can substitute for the occasional missing or inadequate detail by spelling out assembly or whatever; but it is still more efficient to do the drawing properly.

If construction is conventional, the specification could be fairly brief – the contractor could reasonably be expected to know enough to fill in the gaps, and the courts would have a clear enough understanding of normal practice to not require it to be spelt out in the specification. On the other hand, if construction is innovative, the specification will need to be extensive, perhaps even being explicit about methods and buildability issues that are normally the contractor's domain (see Chapter 6).

If quality is to be managed by the use of quality assurance on-site, the specification will need to state explicitly what level of quality is required. The contractor will not generally be able to rely on approval from the architect or engineer. This is because quality assurance, embodying Abrahamson risk allocation principles, is a hands-off exercise in which the contractor is expected to take much more responsibility for the work it does than conventionally.

However, quality is still difficult to quantify. For trades such as roads and landscape, specifiers seem culturally unable to avoid intervention with a large part still being specified "to approval". This is not workable under quality assurance, where the contractor carries out most of the inspections and tests.

Traditional quality control, on the other hand, is all about approvals by the contract administrator (of work executed, of samples and prototypes, of shop drawings, and so on) and so the specification can be less explicit on quality. Something needs to be said on the approval processes themselves, however i.e. inspections, tests, samples and the like. As stated in Chapter 2, the specification must address the means of demonstrating compliance.

If the documentors are not commissioned to follow through on-site it could be argued that the documentation generally needs to be more thorough, more for the benefit of the contract administrator, who will be unfamiliar with the design intent, than for the contractor.

If the contractor is known to the documentor and/or principal, and is experienced in the sort of work required, the documentation could be less thorough. Partnering as a method of procurement could also lead to less thorough documentation, for similar reasons (Uher 1994).

Documents are prepared to control risk. For risk-prone projects (e.g. time-sensitive ones such as shopping malls due to be open for Christmas, or construction-sensitive ones such as nuclear reactors) documents will need to be more thorough. For principals with limited construction funds, documentation will again need to be more thorough, to avoid any time, cost and quality blowouts that may jeopardise the project. On the other hand, if the principal is prepared to accept considerable risk and/or encourage contractor innovation (e.g. the project is experimental, and/or the contractor may have something useful to offer that may save time or cost), then the documents can be more open.

Where high quality work is required it is fair to say that generally a more thorough (and probably longer) specification is needed. This may apply to the whole project or just to a critical part of it. Similarly, if the project is technically simple but repetitive (e.g. town houses) it could also be argued that the specification will need to be more thorough, simply to contain the risk-magnification effect arising from 40 units rather than one. Making prototypes, to discover and resolve problems on-site and then modifying the drawings and specification, is worth considering in these circumstances.

The principal may be particularly fussy about an element of the project and it would be fair to reflect those concerns in a more thorough specification. It is not reasonable to overspecify e.g. requiring a higher level of product quality than is strictly needed, because you are fussy. It must be remembered that the specifier acts on the principal's, not the specifier's, behalf. Mind you, your fussiness may be the very reason you were commissioned in the first place. Some specifiers unintentionally give the impression of being fussy by leaving in marginally relevant material e.g. carpet test schedules for a housing project. If in doubt, leave it out! You have probably no intention of enforcing such material in any case.

For small projects in particular, if the drawings are heavily annotated, the separate specification will have less to say. It may not even exist. Heavy annotation of drawings is also common on larger projects, particularly with the increasing use of computer-aided drafting, for housing projects, and for certain design disciplines. For example, structural engineers commonly include a sheet of standard specification notes (a number of which are invariably irrelevant, out of date and/or at odds with the specification proper) as the first sheet in the drawings set.[10]

In effect, the specification is split between the drawings and the written document. This split conflicts with the principle that something should be described once, and in the right place. For quality matters (for example), the right place is in the specification, not the drawings.

Heavy annotation is really not a good idea. Annotated drawings may appear so complete that the contractor (and tenderers, estimators, contract administrators) may fail to see the relevance of the specification proper, and so not use it. That this happens is perhaps reflected in the DPWS (NSW) decision, in 1995, to specially audit projects to ensure that materials supplied comply fully with specifications. These audits really should not be needed. Contract administrators should routinely follow through on the specifications, whether by proprietary item, standard, test method or whatever. This is not to suggest that annotation of drawings is not required. It is, for full coordination with the specification, and to clarify assembly, extent and position.

10. One argument given to support this practice is that the specification is trashed after completion, whereas the drawings and this first sheet can be microfilmed and archived together. Documentation retention is discussed in Chapter 7.

Pareto principle

As suggested earlier in the chapter, a law of diminishing returns (the Pareto principle) applies to the writing of a specification. It applies as much to the size and thoroughness of the specification itself as it does to the extent to which it has been researched. A lot of research may end up condensed into a few lines in the final document, so there is no direct correlation between size and effort. In fact it is often true to say that the bigger the specification the less effort has been made, since a lot of extra effort is needed to edit the specification thoroughly, making it easier to use. This saves time and confusion later, so editing is a good investment.

Big specifications are almost invariably bad reads. One has to understand the distinction between complexity and chaos. A careful balance between overt size and hidden effort is needed. Unfortunately, few specifiers go to the trouble of editing the specification thoroughly. In particular, architects rarely edit services material, something that is true in the USA too (Heineman 1999a; 1999b). This is discussed more fully in Chapter 7.

Structure – overall

The traditional method of breaking the specification down into worksections is to follow the 'trade' or working sequences on-site. Most master specification systems, for example, follow this principle, while differing on the details. Most rely on classification systems to determine the detailed sequence (see Chapters 2 and 3).

In general, specifications broadly consist of three parts (Figure 4.5):

- Preliminaries.
- Trade or worksections.
- Multi-trade or work schedules.

Standardised contents for what are known in Australia as the Preliminaries are somewhat disputed. Generally, it is agreed that this worksection contains material of interest mostly to the contractor, and not to subcontractors. There are several candidates for what might be included (see also RAIA AN14.03.103 [1999]):

- Contractor's preliminaries e.g. site establishment such as secure storage, hoisting, scaffolding.
- Principal's preliminaries e.g. temporary facilities and services and permanent works components provided by the principal, progress photographs, temporary facilities for the use of the principal.
- Principal's policies e.g. no smoking, employment of apprentices.[11]
- Matters prompted via the conditions of contract. Examples from AS 4000 (1997) include use of second-hand materials (clause 29.1), quality assurance requirements (clause 29.2) and order of work and construction program (clause 32). Examples from JCC include conditions for work on adjoining properties (clause 3.08.03), facilities required by the clerk of works (clause 5.09.02) and evidence of authority approvals sought by the principal (clause 6.05.01) (Gelder 1993b).

11. RAIA Advisory Note AN15.01.603 (2000) suggests that this topic be dealt with in tendering material.

- Special conditions of contract i.e. project-specific variations to standard forms. The RAIA and APCC make recommendations in this area.[12]
- Additional 'commercial' clauses e.g. sales tax, measurement of work, payment for unfixed materials, content of progress claims, valuing work under provisional sums.
- General technical material, relevant to all or most trade sections i.e. to subcontractors and suppliers. This includes rules for quality control procedures and for interpretation of terms used in the worksections (e.g. 'submit').
- Multi-trade or work schedules e.g. door schedule. Standen (1995) suggests that these be included in the preliminaries. This material is of interest to subcontractors and suppliers.

NATSPEC includes only some of these topics in its *Preliminaries* material, which is contract-specific:

- Principal's preliminaries.
- Matters prompted via the conditions of contract, where they can be dealt with proactively and where they are not to be dealt with in the annexures or appendices to the contract.

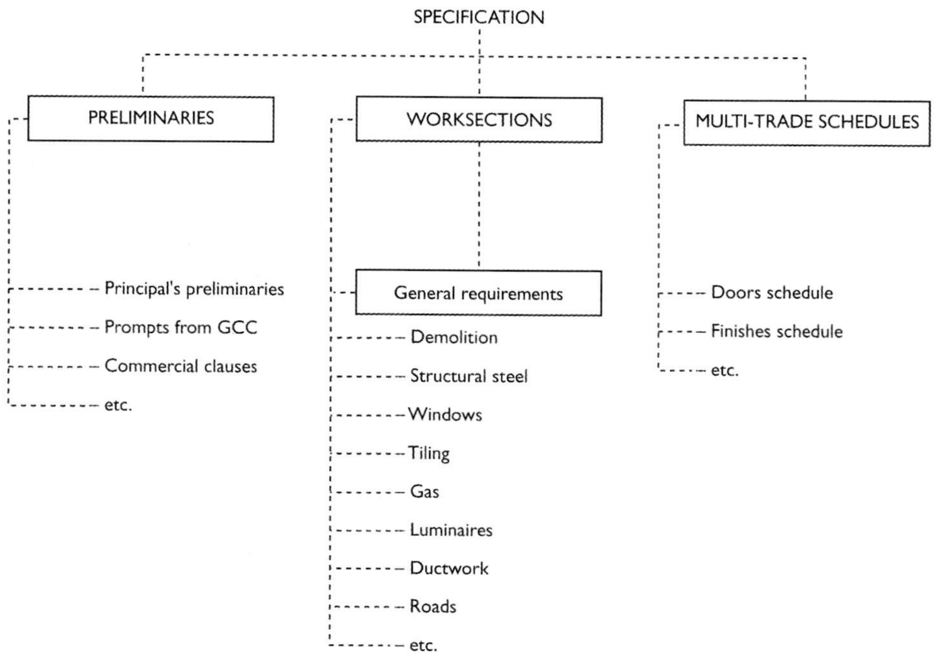

Figure 4.5: Structure of the specification

12. *For example, recommendations are made in the RAIA Advisory Note AN15.02.102 (2000).*

- Additional 'commercial' clauses.

The remaining candidate topics are dealt with elsewhere in NATSPEC, or not at all. Contractor's preliminaries are generally the contractor's business. But they are important for nominated subcontract work, and for multiple-prime contracting (discussed in Chapter 5), and for this reason NATSPEC has a worksection called *Package definitions*.

On principal's policies, NATSPEC has followed the RAIA's advice, providing prompts but no text. Special conditions of contract and annexures are not included. Unlike the specification, they address both the principal and the contractor. They belong with the contract proper. General technical material, of interest to subcontractors, is in a separate worksection (called *General requirements*). Even so, specifiers often include some of these topics in *Preliminaries*, particularly *General requirements*-type material. Finally, NATSPEC provides *Multi-trade schedules*, being of interest to subcontractors, in a worksection of that name.

On matters prompted via the conditions of contract, there are two things to bear in mind: the principle of saying it once and in the right place, and the inconsistent coverage between general conditions of contract. For example, AS 4000 clause 27 and C21 clause 59 mention cleaning. Does this mean that all requirements for cleaning should be covered in the contract (as special conditions), rather than in the specification?

Half breaking the rule, NATSPEC puts other requirements for cleaning into the specification. The contract is not the right place for this topic; cleaning forms part of the description of the works and that is what drawings and specifications are for. Anyway, other contracts (e.g. JCC, SBW-2) say nothing about cleaning[13], so all requirements would properly be given in the specification in these cases. But is the *Preliminaries* the right spot, or should it be in the trade or worksections? The answer to this depends on whether one is addressing project-wide cleaning (*Preliminaries*) or trade-specific cleaning (e.g. cleaning mortar from glass blocks, in the *Glass blockwork* worksection).

Worksections (trade sections) are of contractual interest to the contractor and of practical interest to subcontractors and suppliers. *NATSPEC BASIC*, for example, uses the following workgroups, worksections and sequence:

Preliminaries and general work

- Tendering
- Preliminaries
- General requirements
- Adhesives, sealants and fasteners
- Metals and prefinishes
- Termite control
- Timber finishes and treatment

13. It would be helpful if those drafting contracts could agree on their general scope.

Site and groundwork

- Demolition
- Site preparation
- Earthwork
- Service trenching

Structure and carcass work

- Concrete formwork
- Concrete reinforcement
- In situ concrete
- Concrete finishes
- Light timber framing
- Timber flooring and decking
- Structural steel
- Light steel framing
- Brick and block construction

Enclosure work

- Insulation and barriers
- Waterproofing
- Roofing
- Cladding
- Doors and hatches
- Windows
- Glazing
- Door and window hardware
- Lining
- Suspended ceilings
- Partitions

Finishing work

- Plastering
- Cementitious toppings
- Tiling
- Floor sanding
- Resilient finishes
- Carpets
- Painting

Commissioning work

- Metal fixtures
- Timber fixtures
- Extinguishers and blankets

Hydraulic installations

- Stormwater
- Wastewater
- Freshwater
- Irrigation
- Fuel gas
- Hose reels

Electrical and mechanical installations

- Domestic electrical and mechanical installations

External work

- Fences and external walls
- Landscaping
- Paving

- Multi-trade schedules

The terms 'trade' or 'work', though widely used, do not accurately describe the content of all the sections in a modern specification. Specification sections describe products (manufactured, fabricated and built) which are somehow related in the project. Traditionally the relationship has been via a trade or type of work, hence the names used. But manufactured products used by many trades are often described separately to ensure consistent specification and to avoid repetition.

Adhesives, sealants and fasteners, and *Metals and prefinishes* are examples of this sort of worksection. Some worksections, such as *Concrete reinforcement*, are only about fabricated products. Some, such as *Concrete finishes* and *Floor sanding*, are only about the built product. But most specification worksections are integrated, and describe manufactured, fabricated (if applicable) and built products.[14] Indeed, worksections might even be about complete buildings, such as ticket booths or kit saunas.

There are other types of worksections. 'Umbrella' worksections govern all sections beneath them in the hierarchy. *General requirements*, for example, governs all the worksections in *NATSPEC BASIC*, as it contains the general technical material mentioned previously.

14. These products are sometimes characterised as input and output, the input being the principal material e.g. brick, and the output being the part e.g. walling, paving. Most worksection titles are generated this way in the UK's CAWS system e.g. Brick walling.

Reference worksections do not relate to the work of any one trade but may apply to several worksections, which accordingly rely on them. *Insulation and barriers* and *Glazing* are examples, in which the different project types of these materials are defined (unlike the manufactured product sections such as *Adhesives, sealants and fasteners*, which are generic). Their integration into the built product is described in other worksections, such as *Roofing*, *Brick and block construction*, and *Partitions* (for *Insulation and barriers*), and *Windows*, *Doors and hatches*, and *Timber fixtures* (for *Glazing*), for example.

The *Preliminaries* section and the worksections themselves, are often polluted by specifiers keen to include inappropriate material, such as items to do with tendering, which is precontractual and should be kept entirely separate. The pollution of worksections is a particular problem in engineering material, partly because the lead specifier and the engineering specifiers do not communicate about the content of preliminaries and conditions of contract. This issue is discussed more fully in Chapter 5.

For design-and-construct procurement, the specification may not describe work and products at all, depending on the point of handover. If handover is early in the design process, for example, the specification may describe facilities, spaces, and parts, which will, in effect, be in the form of a brief. A different approach to structure will be needed. Handover documents prepared somewhere between briefing and full documentation will need a hybrid structure, to be resolved to suit the needs of the project (discussed in Chapter 6).

Multi-trade or work schedules do not belong to any one worksection, combining as they do material from several worksections. Door schedules are one example. These deal with frames, different types of leaf (timber, metal, glass), hardware, electrical work (status indicators, electric locks), grilles, signage and glazing. Finishes schedules are another example. These deal with interior and exterior surfaces and schedule a variety of finishes including carpets, paints, metal coatings, tiles and self-finishes for render. They epitomise the tabular approach to text, also used to good effect in individual worksections.

Multi-trade schedules are best located at the end of the specification, after all of the worksections and before the index, if you are to have one (see later discussion of Transparency).

Sometimes they are located on the drawings (often integrated with a graphical schedule), or prepared as part of the drawings set, because they are often too large for the A4 pages of the specification. Fold-out A3 pages could be used instead. NATSPEC material assumes that these schedules will be simplified by the use of project-specific codes, which may be defined in the relevant worksection e.g. lock types coded under *Door and window hardware*, carpet types coded under *Carpets*. Where necessary (e.g. several finishes on a particular floor or wall) these same codes will also be used on the drawings. For multi-trade items in head contract work e.g. reception desks and doors, you may include specific materials requirements (if any) under the item subheading in each worksection (e.g. define hardware codes), but provide central drawings (assembly) and/or schedules. It is clearly important, for these items especially, to use the same term for a given item throughout the documentation.

Contractually, the worksection sequence does not matter. The sequence of the worksections, or drawings, has no effect on the sequence of work. Wherever a piece of work is specified, or drawn, the contractor is obliged to do it. The traditional approach to sequencing is to follow the sequence on-site, with the caveat that mechanical work is described in one block, electrical in another, structural in another, hydraulic in another, and so on. So the detailed sequence does not correlate to who will be on-site at any one time. The carpenter and electrician may be on-site simultaneously, for example, during 'first fix'. The details of this 'trade' (or work) sequence are not universally agreed. For example, external works could be combined with site preparation, mechanical and hydraulic could be combined, and timber and metals could be separated, as they are in the US MasterFormat.

Other types of sequences could be used e.g. worksections could be sequenced alphabetically, or in letting sequence, or by author (architectural block, structural block, interiors block) like the drawings. However, contractors are used to a trade sequence, and despite the incorrect implication that worksections correspond to subcontracts, it is worth sticking with this sequence.

Worksection scope

How do we decide what goes into worksections? Contractually, it does not matter where requirements are stated, providing they are stated only once. But this sort of attitude would make preparation and use of the specification unnecessarily difficult – one would never know where to look for something. Requirements need to be stated in the right place, and this needs to be self-evident. Principles for worksections content are needed, including general considerations (Gelder 1997b) such as

- use;
- procurement;
- transparency;
- worksection size;
- gaps and overlaps;
- consistency;
- tendering;
- subcontracting; and
- alignment

Use: The specification has at least nine functions: to define precisely what is to be built; to be a design record; to show statutory compliance; to be used in pre-tender estimates, in tendering by the principal and subcontract tendering by the contractor; to be part of the contract; to be used on-site; for help in disputes; for information in facility management; and in project management (see also the Preface).

Every user has its own requirements, often in conflict with those of other users (compare the needs of building control officers and tenderers, for example, as discussed in Chapter 8). They cannot all be satisfied. While specifiers would like to be helpful to users, particularly tenderers and the contractor, they cannot afford to lose sight of the need to produce a contractually sound document that will stand up in court.

The key is ease of preparation. If the specification is easy to prepare, then it is likely to be easy to use. Structuring the specification so that it is unnecessarily difficult to prepare and check will make it error-prone during preparation and checking, error-prone in use, and error-prone in the courts. This will do nobody any favours.

Procurement: The specifier does not know the needs of tenderers and the contractor. Subcontracting practice is a floating target, evolving all the while, and varying across the country, from job to job, and from tenderer to tenderer. On simple projects, for example, all doors may well end up as the responsibility of one subcontractor. On more complex projects, it is likely that fire doors will be supplied and installed by one subcontractor, partition doors by another, and so on (see Table 4.3). At best, the scope of worksections can only be loosely linked to the scope of the actual subcontract packages. Some standard US forms of contract make this explicit such as AIA A201 (1997) at clause 1.2.2.

Specification structure is driven by the logical demands of documentation production, and is the responsibility of the specifier. Subcontract and/or supplier packaging is the responsibility of the contractor, and is geared around its customary approach to subletting, availability of subcontractors and suppliers, project location and type, quality requirements, design requirements, cost and time consequences, process integration, and so on. Nevertheless, many specifiers, particularly in the services arena, seem to believe that their material is addressed direct to the subcontractor and so must be organised exactly to suit subcontract packages.

Package specifications, used in multiple-prime contracting (see Chapter 5), may have several (edited) sections. A doors package may include, say, *Preliminaries*, *General requirements*, *Doors and hatches*, *Door and window hardware* and *Glazing*. It does not have to comprise just one worksection.

Transparency: The specification, like any book, must be navigable. Unlike books, specifications rarely have indexes – content should be obvious from worksection title alone. For example, if there were no insulation worksection, no waterproofing worksection, or no termite control worksection, the reader (including the specifier drafting or reviewing the document) would need to check out several possible locations for this material. This wastes time, and increases the chances of missing something, and of inconsistent and repetitious information.

Worksection size: The worksections must be neither too small, nor too large. Like Goldilocks, we want them to be 'just right'. If they are, the contractor will then have fewer short worksections to amalgamate and long worksections to disassemble. Short worksections necessitate a lot of repetition of standard 'general' material, even if a *General requirements* worksection is used. Long worksections are easy to get lost in, and logical hierarchies within such sections are harder to spot or follow. Worksection length should be optimised in an ideal range, say, between 500 and 2500 words (see Table 4.2).

These issues are important to authors (specifiers) and to contract administrators, too. Let's not make more work for ourselves. We should be working smarter, not harder, especially in these times of tight fees and programs.

Gaps and overlaps: The specification is addressed to one entity, the contractor, so a requirement need only be stated once. Indeed, the same product must not be described more than once – there must be no overlaps. Conversely, items must not be omitted – there must be no gaps in coverage.

This is easily controlled if related items are located together, even if this grouping does not correspond to a subcontract package. A good example is termite control, which may combine several approaches to give full protection, all involving different trades. Building control officers would appreciate it if this was all specified together, and the specifier can readily check that the description is fully integrated. Don't forget that specification preparation is part of the design process. The structure of the specification can make this process easy, or hard.

Advantages	Disadvantages
Do not need to be disassembled as much as large worksections	May want to assemble them into large worksections, but this is easier than disassembly
Better alignment to project requirements inherent in the master specification	Alignment may suit larger projects more than smaller ones
Similar work can be specified in one spot, achieving consistency and avoiding gaps and overlaps	Some work may relate to that described in several other worksections; a single small worksection is unlikely to describe a self-contained piece of work
Easier updating and revision	
Transparent content, in the title	
Alignment with international practice	
Readable chunks of text, neither too long nor too short	

Table 4.2: Advantages and disadvantages of small worksections

Consistency: The same work must be described consistently throughout the specification. This, too, is best achieved if a particular type of work is dealt with in one spot only. Service trenching, for example, should be specified centrally rather than in groundworks and stormwater, wastewater, water supply, gas, cabling, telecommunications and so on. Otherwise, even if there are no gaps and overlaps arising from this (which is highly unlikely) there will probably be differences in the specified backfilling and compaction practice.

If one subcontractor does all the trenching, these differences will lead to total confusion, especially if the trenches are shared by several services, as they often are. This will lead to more time, more instructions, rising costs and potential loss of face for someone. At worst, work will need to be redone to the correct specification.

Tendering: A contractor may well rearrange a specification one way during tendering, and another way for actual subcontracting. Given that tendering is a rushed process, disassembly at this stage may well follow the expedient route of exactly matching specification (and drawing) structure. There is no time to resolve possible conflict, double pricing and so on arising out of a poorly structured specification. In any case the contractor will massage the prices received, with a view to winning the job. It is a crude process, which is not helped by specifiers ignoring most of the principles outlined above, particularly to do with repetition and transparency.

Subcontracting: The contractor will disassemble the specification (and drawings) and will find itself replicating standard requirements (e.g. for fastener standards) in several subcontract packages, just as some drawings are replicated. But it is the contractor's job, and not the specifier's, to prepare the subcontract documentation. Contractors might not do this well, but that is their problem. They need to take the preparation of subcontract documentation more seriously – it is not just a matter of rearranging the material. For example, they may choose to close an open contract specification, or they may choose to stipulate methods to be used, and they must properly define package boundaries.

Specification structure can ease this rearrangement of material. For instance, glass may be separately specified in *Joinery*, *Doors*, *Windows*, *Partitions* and *Miscellaneous fixtures*. But the contractor wants to let some to a toughened glass supplier, some to a mirror supplier, some to a SIG (sealed insulated glazing) unit supplier, bullet-resistant glass to a specialist supplier-installer, and the rest, including installation, to a non-specialist supplier-installer. The contractor will then have to remove bits from five disparate worksections and relocate them to create five totally unrelated packages. This is a lot of fiddly work! It would make the contractor's life easier if all glass were centrally specified in just one worksection. It can then be relatively easily subcontracted in any way the contractor chooses. This has the added benefit of ensuring that some minor glass component, buried in what seems to be a 'non-glass' worksection, is not overlooked in the subcontracting process. It also, incidentally, makes life easier for the specifier and contract administrator.

Alignment: Be clear about the distinction between the demarcation between consultants, the structure of the documents and the structure of the subcontracting arrangements (discussed above and also refer to Figure 4.6). They are unlikely to correspond exactly, nor should they. Generally, the specification should be structured for ease of navigation, and should not be influenced by other project demarcations. It is not necessarily aligned to who is responsible for design – material designed by the architect may be included in the mechanical or electrical sections, and vice versa. Generally, this is the design manager's responsibility. Nor is the specification necessarily aligned to the source document e.g. NATSPEC. The project specification may need to be radically restructured.

Structure – worksections

There are several options for the organisation of worksections (see Table 4.4). The North American SectionFormat breaks each of its worksections down into three parts, or subsections: general, products and execution (see Chapter 3). For many items this forced separation of product and associated execution can be fairly cumbersome. Nevertheless, it helps the contractor as 'supply-only' information is readily isolated. Within these three subsections SectionFormat uses a standard sequence of clauses.

Conversely, the Dutch national master specification system (STABU) breaks worksections down into a library of self-contained clauses corresponding to components, combining general, products and execution material for each component. Clause-to-clause variation may be very minor in this system.

In its newer material, NATSPEC's structure is a little more elaborate than SectionFormat, and is broadly as follows:

- General, including design.
- Quality (management).
- Materials and components (manufactured products).
- Execution, including custom manufacture and fabrication (custom-made products).
- Completion.

Like the SectionFormat system, the NATSPEC system is hierarchical, running from the general to the particular. Therefore, unlike the STABU system, material on a particular component cannot simply be relocated e.g. text on fire doors in the *Doors and hatches* worksection cannot be simply relocated to create a *Fire doors* worksection. Governing material in previous clauses may also be relevant. Unlike current NATSPEC material, the STABU system relies on a relational database, making the modular approach to products taken by STABU feasible, if not essential.

NATSPEC material is not a library system of alternative clauses. Rather, it works by the specifier deselecting unwanted material from a default master. This approach works best if the right starting point (master set) is selected in the first place. Using *NATSPEC*

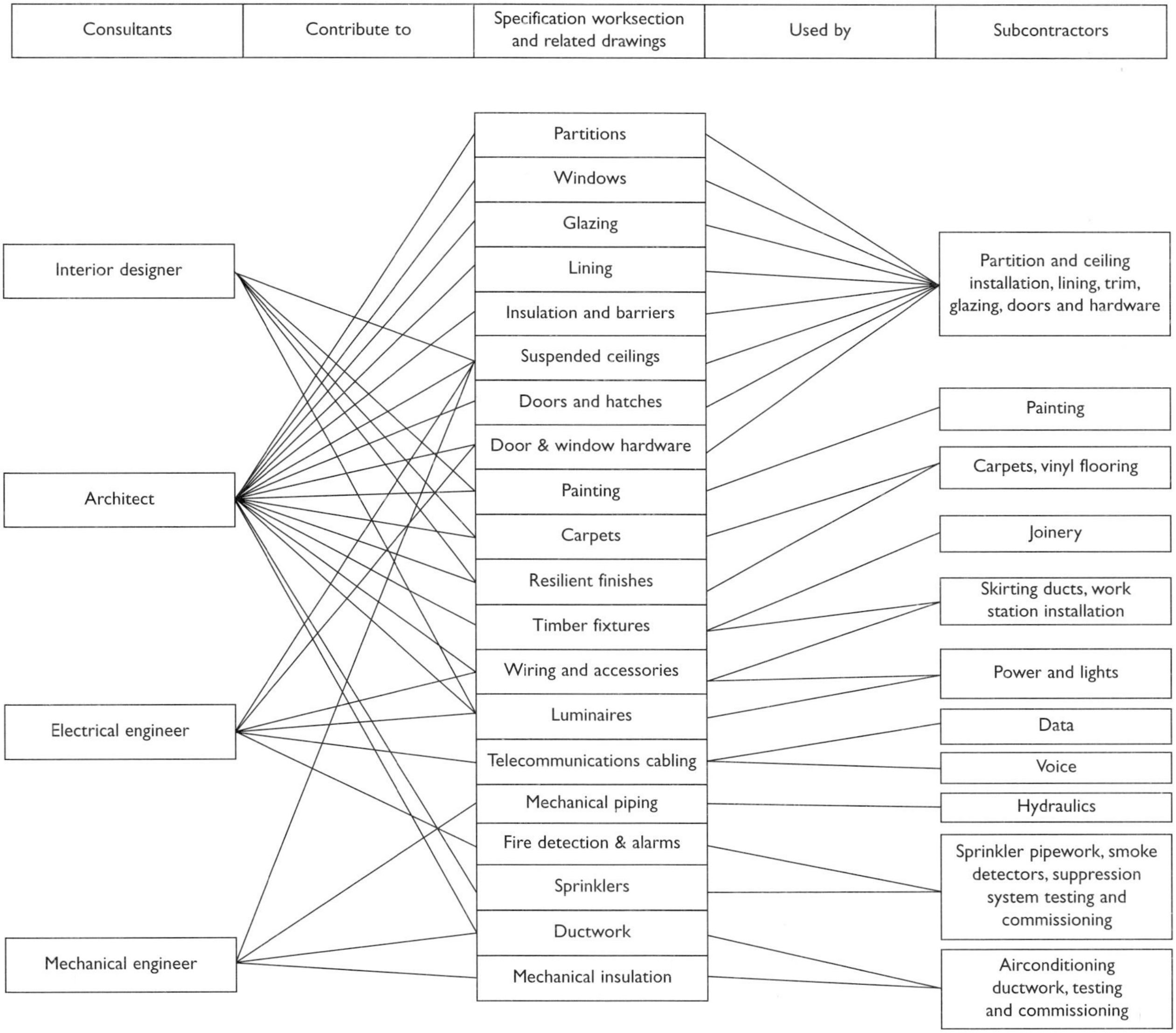

Figure 4.6: Fit-out subcontracting. The structure of the specification has little to do with design or subcontracting. It has more to do with knowing where to find the material. It must be stressed that subcontracting practice varies from contractor to contractor. It is they, not the specifiers, who allocate the work.

Package	Project A	Project B	Project C	Project D
Termite treatment			●	
Precast concrete				●
Precast concrete supply		●		
Laminated timber columns	●			
Sundry carpentry	●			
Roof trusses and fascias			●	
Blockwork				●
Masonry		●		
Masonry supply	●		●	
Masonry laying	●		●	
Waterproofing	●		●	●
Tanking		●	●	
Waterproof membranes				●
Metal roofing			●	
Roofing and plumbing	●	●		●
Sandwich panels	●			
Fire doorsets and steel door frames		●	●	●
Fire doorsets	●			
Metal door frames	●			
Stainless steel fire doorsets	●			
Stainless steel door frames	●			
Sliding and pivot fire doors				●
Non-rated timber door supply	●		●	●
Special doors				●
Door hanging and hardware fixing	●		●	●
Roller shutters	●	●	●	●
Shutters and screens		●		
Windows and frameless glazing				●
Aluminium windows and doors			●	
External glazing	●			
Mirrors	●		●	
Hardware supply	2 packages	●	●	●
Sanitary hardware				●
Timber lining	●			
Fabric lining	●			
Standard/suspended ceilings and packages	●	●	●	●

Package	Project A	Project B	Project C	Project D
Toilet partitions and doors	●		●	●
Shower screens			●	
Operable and acoustic walls	●			●
Terrazzo	●			
Wet plaster and granolithic toppings				●
Rendering	●	●	●	
Applied finishes and rendering			●	
Ceramic/quarry tiling	●			●
Ceramic tile supply			●	
Ceramic tile fix			●	
Stone and marble work				●
Carpet and vinyl			●	
Resilient finishes	●			
Vinyl tile and skirting				●
Carpets	●	●		●
Internal painting				●
External painting				●
Painting	●	●	●	
Fabricated metalwork	2 packages	●	●	3 packages
Metalwork supply items		●		
Sunscreen and louvres	●		●	●
Glazed awnings			●	
Fabric awnings	●			
Fire stair handrails				●
Balustrades	●			
Aluminium handrails and balustrades			●	
Aluminium work		●		
Joinery		●	●	
Fixed furniture	●	●		
Joinery - back of house				●
Seating frames	●			
Seating	●			

Table 4.3: Examples of multiple-prime contracts. These similar Queensland projects (Professional Specifications 1998) were let as a series of separate contracts, and each has a distinct approach to contract package scope.

BUILDING for a single house would be impractical, for example – *NATSPEC DOMESTIC* would be better. The STABU system is a true library system, working by selection of wanted clauses. The process, potentially clumsy given the number of alternatives in certain cases, is facilitated by a user-friendly computer interface. The STABU system also tends to be used only on larger projects.

The consistency of worksection structure within a specification can be a problem where those worksections prepared by other authors (e.g. by manufacturers or by co-consultants) do not follow the same structure as the architectural worksections, for example. This can be sorted out by editing or, better still, by briefing co-specifiers up-front on what is expected. As always, prevention is better than cure.

Summary

Use reliable up-to-date sources. In particular, cull unwanted material, use national master specifications as far as is practicable, and avoid cut-and-paste. All sources should be used critically.

Allow adequate time and adequate fees for research, preparation and review of specifications. They take around 10% of documentation time, for example. Use the right person for the job, someone who is ready, willing and able. Provide training for specifiers.

Develop the specification systematically. Know when to stop (as a guide, use the 80/20 rule), and when to start (not too late). Prepare, and follow, written specification-preparation procedures, including systematic updating.

Make the specification as thorough as is reasonable in the circumstances. Know when to stop (again, use the 80/20 rule as a guide). Minimise drawing annotation.

SECTIONFORMAT	NATSPEC	STABU
GENERAL	GENERAL	GENERAL
	QUALITY	
PRODUCTS Element A Element B	MATERIALS AND COMPONENTS Element A Element B	ELEMENT A General Products Execution
EXECUTION Element A Element B	EXECUTION Element A Element B	ELEMENT B General Products Execution
	COMPLETION	

Table 4.4: Worksection structure – alternatives

Split the *Preliminaries* into contractor-focused material and separate subcontractor-focused material (e.g. NATSPEC *General requirements*). Locate elsewhere i.e. not in the specification, any required contractor's preliminaries, principal's policies, special conditions of contract and contract annexure. Do not include tendering material in the specification.

Structure the worksection component of the specification (in terms of sequence and content) for navigation, ease of reference, and for ease of disassembly, not necessarily for likely subcontracts. Use trade sequence as an overall guide to structure, though this is not at all precise or fixed. Note that the structure of the project specification is not necessarily aligned to the source specifications, nor to consultant responsibilities.

Use non-trade worksections where appropriate e.g. for insulation, glazing or sealants. Separate any multi-trade schedules and locate them all at the end of the specification. Structure worksections internally in a consistent manner, and for ease of disassembly.

5 Contractual issues

Specifiers need skills and expertise in three broad areas: contractual, technical, and editorial. These will be examined in more detail in this and the next two chapters.

Contractual matters are particularly important because of the pivotal role of specifications in the contract documents. Specifications link all the other documents and so must be coordinated thoroughly with them. Specifiers therefore need to understand what is in these other documents even though they may not have prepared them.

This is often a problem for services and structural specifiers who are not always privy to the conditions of contract and the *Preliminaries*. Accordingly, their specifications often cover topics dealt with (better) elsewhere. It can also be a problem for subconsultant specifiers, who may not be working off current drawings.

But first the chapter provides background information on various procurement routes for buildings, and types of contractor. It then looks at the broad issue of coordinating specifications with the conditions of contract, the drawings, the bill of quantities (if any), referenced documents such as standards, and the complex issue of coordination with other contracts. This is followed by a review of the principle of equity and its impact on the specification.

Procurement

The nature of the contract documents is fundamentally affected by the chosen method of procurement i.e. how the building is to be acquired. For our purposes there are essentially two paths for contracting – single-prime and multiple-prime,[1] and two types of contractor – construct only and design and construct. Needless to say, thanks to human ingenuity, there are myriad variations on these themes.[2]

Single-prime contracting, the traditional procurement path, is also known as head, lump-sum or general contracting. The entire project is tendered in one shot, with the

1. The terminology, which is American, is intended to clarify the distinction between (prime) contracting and subcontracting. Alternative terms are a little less helpful in this respect.

2. Procurement practices vary considerably from country to country. A European survey (Mindt 1995) found that multiple-prime (divided) contracting was the norm in Switzerland (90% of projects) and France (80%), common in Finland (45%) and rare in Sweden (5%). Single-prime (general) contracting was the norm in the UK and Sweden (75%), and common in the Netherlands (50%) and Finland (35%). It was uncommon in Switzerland (10%). Design-and-construct contracting was common in the Netherlands (50%), fairly common in France (15%), Sweden and Finland (20%), and uncommon in the UK (8%).

services elements etc. included under the single-prime contractor. All the works description documents, including the specification, are addressed to this one contractor. A standard form of contract may be used, such as AS 4000 (1997) or JCC (1994a).

The resolution of subcontracting and supply arrangements, including tendering and site coordination, becomes entirely this contractor's responsibility. The principal and contract administrator are not privy to these processes. This contractor is in direct contract with the principal.

Multiple-prime contracting is also known as package contracting. With this form of contracting essentially there are many single-prime contractors on-site, all bound by standard forms of contract such as AS 4000 or JCC. They are tendered separately on behalf of the principal (usually by a specialist agent such as a construction manager), often in a sequence to suit off-site activities.

This means they will be documented out of sequence from the designer's point of view. Services work in particular may be under one or more multiple-prime contractors. Each contractor will work from a self-contained set of documents that must be fully coordinated by the principal's team. Conditions of contract should be the same, and interfacing (e.g. provision of temporary facilities and services, coordination in time and attendance) between contractors must be resolved. Each contractor is in direct contract with the principal. Each contractor is responsible for its own subcontracting and supply arrangements.

The general rule that the principal's team is not involved in subcontracting partly falls down when there are nominated subcontractors.[3] These may be used in single or multiple prime procurement, and may be construct only or design and construct (see below).

The RAIA does not recommend the use of nominated subcontractors, preferring alternatives (see RAIA Advisory Note AN16.10.100 [1995c]). One reason for this is that the principal is responsible for interfacing between the nominated subcontractor and the relevant (prime) contractor. For the sort of projects where nominated subcontractors are used there is usually no project manager, so this responsibility falls to the design team, which is generally inexperienced in this sort of thing. The form of subcontract should be stipulated, to ensure compatibility with the prime contract (this is not possible for 'ordinary', or 'domestic', subcontractors, because the contractor chooses the subcontract). Nominated subcontracting is falling out of favour.

3. The administration and definition of nominated (or selected) subcontracts is covered at length in AS 4000 (1997) and JCC (1994a), but not in SBW-2, which makes no provision for their use ('identified' subcontractors are the SBW-2 equivalent – a Clayton's nominated subcontract?).

A construct-only contract is the traditional type of contract. With this form of contract the design team is commissioned to prepare 'full' prescriptive documentation where the specification describes products (manufactured, fabricated and built). The contractor simply builds to that documentation with little or no need to make design decisions – the design and documentation team has made them all. The standard forms of contract such as AS 4000 and JCC assume this approach, and make allowance for minor omissions or silences in the documents (see discussion of general defaults, below).

The last type of contract is the design-and-construct contract. With this form of contract the contractor tenders on performance documentation[4] instead of fully prescriptive documentation. The specification may describe facilities, spaces, parts/systems or products, depending on the point of handover.[5] The extent of contractor design, then, also depends on the point of handover (to the contractor).

The contractor is commonly responsible for full design, documentation and construction. Variations include design development, document and construct, in which sketch designs are provided, and document and construct, in which a fully resolved design is provided (Figure 5.1).

Under multiple-prime contracting any one or more of the contracts may be a design-and-construct contract. Services (i.e. hydraulic, mechanical, electrical) contracts and curtain wall contracts are commonly design and construct. Non-standard forms of contract, such as AS 4300 (1995) with the subcontract AS 4303 (1995), will be used for design and construct.

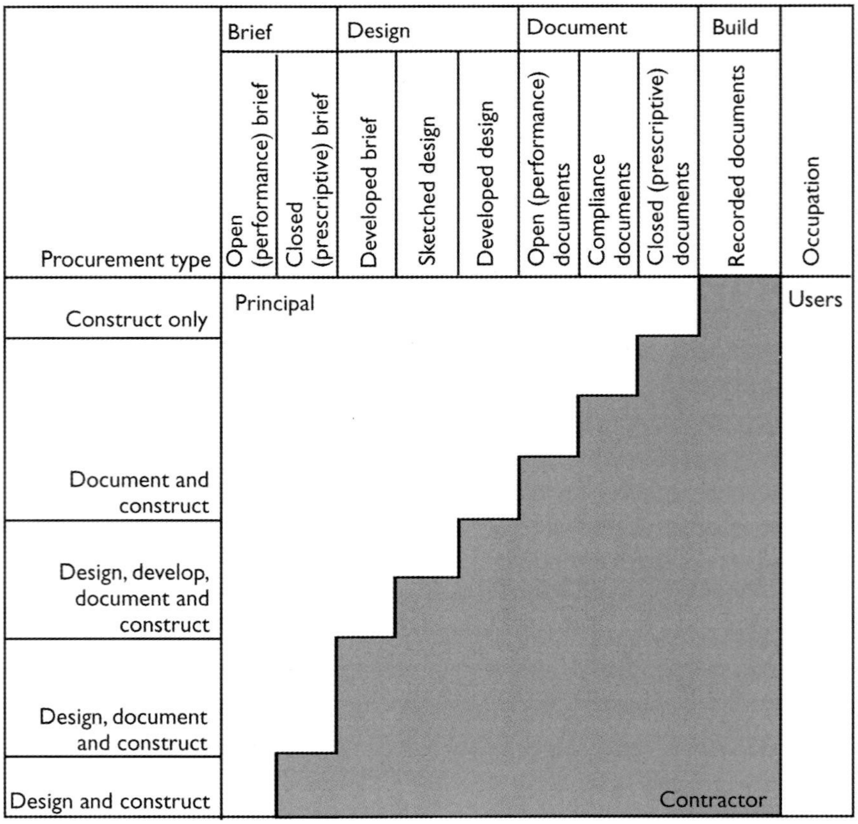

Figure 5.1: Procurement routes and participants

4. *Prescriptive and performance-based specifications are discussed in Chapter 6.*

5. *Prescriptive specification of facilities, spaces, or parts/systems is also possible, but is rare.*

Fast-tracking

Certain combinations of single-prime and multiple-prime contracting with construct-only and design-and-construct procurement lend themselves to fast-tracking, in which tendering of different elements is staggered to ensure the fastest possible time. Other combinations, as shown in Table 5.1, do not.

In fast-track procurement, the various products in the building are documented out of the usual sequence, to allow for the different times taken for materials procurement and manufacture, and for the different points in the program where these products are required on-site. For example, materials procurement and manufacture might take anything from five weeks for ductwork, to 20 weeks for custom-made panellised cladding. For some products, such as plastering, there is no manufacture as such, and for others, such as escalators and lifts, there is no materials procurement phase to allow for. Preparation and approval of shop drawings might also need to be provided for (anon 1995).

Coordination with conditions of contract

The key principle, as already mentioned, is simply: say it once, and in the right place. Too often, contractual material is repeated in the specification (Gelder 1994d). Conditions of contract are the right place for the 'rules of the game'. Specifications do not need to repeat this material. Indeed to do so can be dangerous as it can create conflicts and ambiguities.

Conditions of contract influence the other contract documents in a number of ways, as follows:

- Order of precedence for documents.
- General default requirements for materials, workmanship and contractor responsibilities.
- Definitions.
- Prompts in the conditions of contract, discussed under *Structure - overall* in the previous chapter.
- Commercial requirements.
- Construction methods.
- Contractor design.

It is probably safe to say that most specifiers barely consider the impact that the conditions of contract have on the specification, except in the preparation of the *Preliminaries* material. Here, they seem to understand that there is a link, even if they are not quite sure what it is.

Procurement type	Single-prime		Multiple-prime	
	One-track	Fast-track	One-track	Fast-track
Construct only	Yes	No	No	Yes
Design and construct	No	Yes	No	Yes

Table 5.1: Procurement possibilities

Table 5.2 provides a handy cross-reference to contract clauses on most of the topics discussed below. Many of these clauses are given in Appendix 3, which you might wish to refer to throughout this chapter.

The relative status of the contract documents fundamentally influences their preparation and use. There are two approaches.

In one, the contract sets an order of precedence, as in JCC clause 2.02 in which some documents (still to be read as complementary) override others. In this case the conditions of contract override the specification, which, in turn, overrides the drawings if there is a conflict. Conflicts are therefore self-resolving, but not always in the way you would intend e.g. the courts will turn to the specification before they look at the drawings, at odds with the normal documentation-team emphasis on the drawings. Nevertheless, this

Topic	SBW-2	JCC-C	CIC-1	AS 4000	AS 4300	PC-1	C21
Discrepancies	2.2, 2.3.1	2.02	B1, B2, B4, B5	8.1	8.1	6.10	31.1, 31.2, 31.3
Quality of materials and work	6.6	6.08	M21	29.1	30.1	9.1	37.1
Inferred items	-	6.01	-	-	-	8.2, 9.1	10.12, 10.13
Materials, labour and plant	-	6.02	-	28	29.1	8.2	37.1
Cost of testing	6.6.3	6.08.05	G19	30.7	31.7	9.4, 9.5	38.1, 38.2, 38.3
Legislation	6.3	6.05	M21	11.1	14.1	8.3	14, 42
In writing	5.3	5.04	D13, F17, G9, G13, G17, G21, G27, H5, H15, J13, J21	20	23	16.7	3
Construction methods	6.1	6.03	G1	-	-	-	-
Programming	-	-	G4	32	33	10.2	11
Contractor design	-	-	-	-	4.1(d)	6.2(a)	32.1
Fitness for purpose	-	-	-	-	4.1(e)	6.6	22.1, 37.2
Environment protection	-	-	-	-	-	8.20	13.1, 13.2

Table 5.2: Key clauses for specifiers in general conditions of contract

predetermined precedence offers an easy way out, and a degree of certainty, but this can perhaps be overridden in a special condition of contract, which would reduce its value. Overall, it is worth remembering that courts generally prefer written documents to drawings.

In the other approach, no order of precedence is set, as in AS 4000 clause 8.1, which states that any discrepancies must be notified. The uncertainty created, with conflicts being resolved on a case-by-case basis, is perhaps undesirable at tender. On the other hand, this encourages removal of contradictions before going to tender, which is surely a good idea. And every conflict has to be resolved by making an active decision, so the solution chosen will be the best in the circumstances (and if the conflict is resolved too late, expediting construction may outweigh everything else).

Most conditions of contract contain catch-all clauses such as those dealing with materials and workmanship, as in AS 4000 clause 29.1 and JCC clauses 6.08.01 and 6.08.02. JCC clause 6.01 deals with 'silence' in the documents such that the onus is on the contractor to use suitable materials and methods in these circumstances (this requirement is an "implied term" in simple contracts).

Other catch-all clauses include those dealing with provision by the contractor of all labour and material required for execution of the works (see AS 4000 clause 28 and JCC clause 6.02) and compliance of contractors with authorities' requirements, including giving of notices and paying of fees (see AS 4000 clause 11.1 and JCC clause 6.05). There is no sense in specifications repeating any of these requirements.

But, for those clauses dealing with materials and workmanship, repetition cannot be helped. This is exactly what the specification is about, especially given our key principle that the right place for everything on this subject is the specification, not the contract.

That this makes sense becomes apparent when extreme clauses in contracts are examined. One example is CIC-1 (1998e) clause M21, which ambiguously states: "Where Australian Standards apply the contractor must comply with them, unless required by the contract documents to comply with a higher standard". It would seem that this only refers to those standards cited in legislation (and, if not, the requirement is entirely unworkable), but does this include all the deemed-to-satisfy standards in the BCA? The specification may well have chosen other ways of satisfying the BCA, and these may not be to a "higher standard", but to an equivalent standard.

More extreme, and almost entirely unworkable, is PC-1 clause 9.1. This covers silence in the documents by relying on a reference development for materials and workmanship. This is difficult to manage without demolishing the reference development to see how it was put together, or without accessing a copy of its record specification (if one was made). The clause also requires general compliance with "all relevant standards of Standards Australia" to the extent they are not inconsistent with the contract. As we will see later, this is unenforceable and meaningless. C21 (DPWS 1999a) clause 37.1.5 is even worse in this respect, since it does not have the inconsistency exemption, ensuring there will be conflict between the contract provisions and standards.

Even simple, and common, contractual requirements such as those in AS 4000 clause 29.1 ("suitable new materials and proper and tradesmanlike workmanship") are unenforceable. What is suitable? Different architects and engineers would disagree on this one. Why new, and what is new anyway? Many materials on the market are wholly or partly recycled these days, and for many this is the preferred default. What is proper and workmanlike? We know that 'workmen' routinely omit wall ties if left to their own devices, for example but is this acceptable?

There is no sense in contracts tackling materials and workmanship at all. They must trust the specification, just as they trust the drawings, but the specification must earn that trust.

The conditions of contract (generally in an appendix or similar), and not the specification, are the place to define the works in outline (strictly, redundant, as this is repetition), and the contract documents. For example, see JCC Appendix Items B.4 (works) and B.5 (documents), and C21 Contract Information items 3 (works) and 40 (documents).

Standen (1995) argues that the short description of the works should be located in the *Preliminaries*, because no-one sees the contract until after it has been signed, and then its circulation is restricted. This is not correct. It is usual for the appendices etc. to be completed as far as possible pre-tender, and supplied to the tenderers. Often these appendices are bound into the same volume as the specification, for convenience. This does not mean that they are a part of the specification. They are not; they are the project-specific part of a two-part contract.

It is odd that "days" are differently defined in AS 4000 (calendar days, clause 1) and JCC (working days, clause 9.13), affecting notice for inspections and the like. NATSPEC generally assumes working days. Perhaps this should be stated in the project *Preliminaries*.

A number of topics relating to commercial requirements are covered in one form of contract but not in another. For example, AS 4000 deals with cleaning (clause 27), with payment for unfixed materials (clause 37.3), the number of copies of documents to be supplied by the contractor (clause 8.3), publicity and classified documents (clauses 8.5 and 8.6) and survey marks (clauses 26.1 and 26.3). JCC, meanwhile, does not deal directly with any of these matters (though in clause 10.04.02 it anticipates that off-site materials will be dealt with elsewhere). But JCC also provides for provisional contracts (Appendix Item D) and work by separate contractors (Appendix Item F), whereas AS 4000 mentions neither of these topics.

It is up to the specifier to decide whether the *Preliminaries*, or *Tendering*, or special conditions of contract are the best place to deal with such topics. NATSPEC has a selection of them in its *Preliminaries* worksections. Ideally, anything that comprises a written description of the works (text on cleaning falls into this category, for example) should be in the specification only (remember, say it once and in the right place). Splitting it between the contract and the specification is unhelpful at best.

Commercial requirements are sometimes scattered through the specification, rather than pulled together into, say, *Preliminaries*. This is particularly so for provisional sums, for items supplied by the principal, and for lists of pre-approved subcontractors. Ideally, these would be located in the one spot.

Construction methods (the means) are generally the contractor's responsibility while products (the ends) are the specifier's. This is implicit under quality assurance, which attempts to make the contractor responsible for its actions. Some forms of contract make this explicit. For example, JCC at clause 6.03 defines the contractor's responsibilities as including methods and coordination. 'How to do it' clauses and predetermination of subcontract boundaries are effectively excluded by such clauses.

Corresponding American forms of contract are also explicit on this issue. SBW-2 (clause 6.1) is more forthright, but not AS 4000, surprisingly (the superintendent can give directions regarding sequence of work). To minimise liability under occupational health and safety legislation, DPWS (NSW), which uses C21, used to advise specifiers to avoid specifying that the contract administrator will exercise control over temporary work, work methods or site activities.

It seems much simpler to avoid specifying methods of construction altogether (the next chapter gives other reasons for doing so). A special case is programming of the work, in which the principal may have requirements or constraints that the contractor needs to take into account. Some contracts require a program to be submitted while others do not, some make it contractual while others do not (e.g. CIC-1).

Even in construct-only contracts, the contractor is often required to do design work, often significant i.e. the contracts are in reality partial design and construct. Contractor-design is consequent upon performance specification. This can be extensive in a so-called construct-only specification, from the trivial (use of words like 'sufficient' or 'adequate'), to the moderate (concrete specified by strength), to the significant (curtain walls and mechanical services are often substantially performance-specified).

The problem is that most construct-only forms of contract are not designed for this. Of those scheduled in Figure 5.3, only PC-1 and C21 address the subject (AS 4300 is intended for full design and construct). Issues that should be addressed include fitness for purpose (work that is both designed and built by the contractor can be warranted as fit-for-purpose by the contractor), and the sort of material one finds in client-architect agreements, including professional indemnity insurance, the production and submission of design and construction documents, and intellectual property rights.

Interestingly, the UK construct-only contract JCT's PWQ (1998) provides for a staged response to contractor design. Clause 42 covers design of simple, small and or visually unimportant items such as trussed rafters, or simple lighting installations (roughly speaking, NATSPEC subsections or smaller).[6] The Contractor's Designed Portion Supplement is intended to cover complex, substantial and/or visually important items, proposals for which must be examined at tender, such as curtain walls and complex airconditioning installations (roughly speaking, NATSPEC worksections). Full contractor design (design and construct) is dealt with in a separate contract, JCT's WCD (1998). This staged, or optional extra, approach is not likely in the near future for Australian construct-only contracts. The RAIA, for example, is antipathetic to the whole idea of design and construct (Haslem 1996).

It might be said that principals probably expect the commissioned designers to be entirely responsible for design under construct only, and are not aware that this responsibility is, in effect, split with the contractor. That is, the tender sum includes design fees, yet the contractor is not a subconsultant to the designers for this design work. Just who is responsible for the performance of the design? Has the principal paid twice for it? Is the principal missing out on the opportunity for warranties of fitness for purpose for this hidden contractor-designed and built work? These issues are discussed a bit more fully in Minogue and Klein (1997).

On this last point, contractor-design also arises out of silence in the documents. Where contractors fill gaps in the specifications, they are likely to be responsible for fitness for purpose (Bingham 1995). Not surprisingly, while contractors seem happy to work to full design-and-construct contracts, they do not seem to welcome responsibility for contractor-design under construct-only contracts. *No Dispute* (NPWC/NBCC 1990) states that specifying standards (increasingly performance-based) placed

.. an onerous burden on contractors which may unduly involve them in design and material selection decisions, which are more appropriately the responsibility of the specification writer.

In the USA we find similar concerns (anon 1998):

[AIA A201] Section 3.2.10, Contractor Provided Design Services, ... was one of the sections causing most dissent among contractors. It appears to some to take a great deal of responsibility from the owner/architect, because it has the potential to shift the design responsibility to the contractor.

Changing the contracts

We have talked about modifying specification practice to accord with the content of contracts. This is not always an appropriate course of action, and certainly is not for the specification of quality of materials and workmanship. Indeed, contracts can create problems for the description of project quality by pre-empting the specification. This is a reflection of the quality of project specifications. Principal-written conditions of contract in particular, such as PC-1 and C21, contain the sort of extreme quality catch-alls we have noted because

6. JCT PWQ (1998), clause 42.1, requires that any performance-specified work covered by the clause be identified in the Appendix. This listing would be lengthy, tedious to prepare, error-prone and difficult to use. The American AIA A201 (1997), clause 3.12.10, is not so fussy. It says, "The Contractor shall not be required to provide professional services which constitute the practice of architecture or engineering unless such services are specifically required by the Contract Documents for a portion of the work ...". Probably simply stating performance requirements in the specification is sufficient to 'specifically require' such services.

the principals have been let down by poor specifications in the past. Other quality catch-alls, which are included in consensus forms, may well have been principal-driven too, with other moderating influences operating through the consensus process.

The alternative is to modify contracts to accord with good specification practice. But project-specific changes to contracts should only be made with legal advice. Generally, one has to work with them as published. So two things must happen side-by-side: we must improve the quality of project specifications and we must work to change the content of standard forms of contract.

Coordination with drawings

The coordination of specifications with drawings can be reduced to three main practices: use of consistent terminology, non-repetition and cross-referencing.

The main difficulty is that notes on drawings are often too verbose and would be better in the specification. Problems arise largely because the specification is left till last when the drawings are virtually complete, warts and all, and the specifier has had no opportunity to influence them. This is compounded because the project architect and engineers rarely vet the two side-by-side. On small projects it is relatively simple to coordinate specifications and drawings, but such coordination on large projects must be programmed and ongoing.

Use consistent terminology to avoid confusion. If the drawings refer to 'rooflight', it is reasonable for the contractor to look for this keyword in the specification, and not for synonyms such as 'skylight' or 'roof-window'. You may even have called something else in the project a 'skylight', making it even more confusing to use these two supposedly complementary documents together. Standardised building terminology is listed in a number of publications (for instance, Milton 1992 and 1994; Standen 1993).

A common example of repetition is where finishes are both scheduled in a convenient table in the specification and annotated on the drawings. To avoid repetition, simply describe each of a component's attributes once – most components will require both drawings and text to describe requirements fully.

These two descriptive elements should complement each other, with no overlaps in the description, and no gaps. Cross-referencing will be essential, in which case use identical terminology in the drawings and the specification so the documents can be read side-by-side. For example, use frame, subframe, upper sash, lower sash, mullion A, mullion B and so on for the elements of a window.

Cross-referencing of particular drawings in the specification either should not be done at all, or done properly. To simply state "as detailed" is unhelpful (where is the detail?) and may be incorrect (the detail may not have been drawn after all, and it can be difficult to check this). If it has been detailed, then it is redundant to say so as all documents are read together. It is better to state "as shown in drawing A5 detail 16", for example, which lets the final reviewer and all later users know where to look. The reverse – cross-referencing the specification from the drawings – should also be specific (e.g. "as specified under Partitions – Plenum baffles"), or not done at all. Do not simply state "as specified".

Alongside this, you have to decide what will be specified and what will be drawn. It is mainly a question of efficiency, so make this decision early on. No-one wants to waste time generating a long written description when a drawing would be better, and no-one wants to waste time annotating a drawing with finishes when a simple schedule would be more helpful. And it is dangerous to provide what should be the same information in both places. Conflicts are inevitable. This particularly applies to annotation of drawings, which should be

- minimal; and
- generic.

The specification is the best place for particulars such as proprietary name, thickness and colour. If this information is on the drawings for, say, a flashing, a number of questions arise. Does this information get repeated every time the flashing is drawn and noted? If not, how do you decide which drawing of the flashing is to have the full note and which are to have minimal notes? And will the contractor easily be able to find this single full note? If changes are to be made, will you be able to locate every note to change them all? This is nearly impossible on larger projects (computer-aided drafting may make it easier). Use the specification for this information – that's what it is for.

Coordination with bills of quantities

Unlike the drawings and the specification, which together fully describe the work required, bills of quantities are derivative documents that add nothing to the description (Figure 5.2 and Chapter 7). They do, however, re-present the information in a format that is particularly useful to estimators. Estimators, whether on the documentation team or among the tenderers, should still cross-refer to the specification and drawings for a full understanding of what is required – the cost of 50 m^3 of concrete will vary between an exposed curvilinear opera house roof or a flat concealed floor slab, for example.

Yet when bills of quantities are provided, estimators often do not refer to the drawings and/or the specification, preferring to work off the bills alone. The now defunct Australian Federation of Construction Contractors had this to say on the subject:

Clients may expect that ... tenderers would price the drawings and specification, but the reality of the tendering process is that usually, for the sake of practicality and expediency, the Bill is necessarily used as the basis for pricing (AFCC submission on bills of quantities, undated copy).

Contractors commonly provide subcontract tenderers with the relevant bills only, so they do not even get a chance to look at the other 'parent' documents. That is, estimators and contractors seem to assume that, for their purposes, the drawings and specification have nothing to say that is not said more conveniently in the bills. This is not the case at all, of course. Lack of cross-referencing in the bills does not help. There is no prompting of estimators to look elsewhere for more information.

Cross-referencing from the drawings or specification to the bills is not required (especially if the bills are non-contractual) and is rarely, if ever, done. The exception is perhaps the specified bill, in which the specification text and corresponding quantities are grouped together in the one document. Specified bills are not common in Australia, and are generally prepared by quantity surveyors alone rather than both architects and quantity surveyors. This is not a good idea because preparation of such a document entails design decision making, and quantity surveyors are not designers. The quantity surveyors do use notes provided by the architect and engineers, who should be given the opportunity to review the final document.

Coordination with referenced documents

Specifications widely reference standard documents published by others, most notably Australian standards. Coordination with such documents has to be one-way and, as will be seen, sometimes becomes quite complicated, to the point where the specifier may wonder if it would not have been better to start from scratch. Technical coordination is dealt with in the next chapter; here we concentrate on contractual matters.

In some respects standards are less of a problem than they might appear. For example, many project specifications, at least in the private sector, specify by proprietary item for handbasins, tiles, bricks and so on. The relevant standards are not cited. On the other hand, a number of public sector clients prefer specifications that reference standards, making the issues flagged below significant.

The difficulty is that not many specifiers have – or are even familiar with – all the standards they use. Standen (1995) lambasts specifiers who do not read the standards they

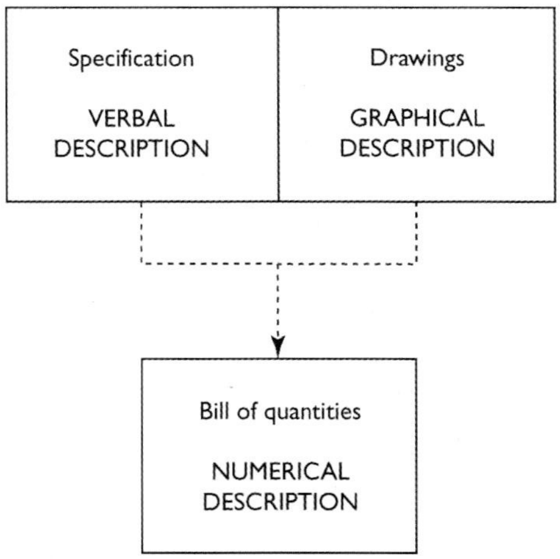

Figure 5.2: Bills of quantities as derivative documents

use, but then suggests that offices do not really need all of them. Likewise, RAIA AN05.01.100 (1996c) advises that architects should be familiar with the standards to which they refer in specifications, but notes that few architectural offices maintain a complete library of all standards covering architectural work (an understatement), and offers a shortlist of core standards (see Chapter 2).

This ambivalence is also found in other RAIA advice. On what might be called the pro-standards side, the RAIA recommends that architects have a general understanding of the use of standards, be familiar with standards relating to the practice of architecture (including contracts), and reference relevant standards for construction techniques and installation of components and services (RAIA AN05.01.001 1996b).

Architects should know enough about standards cited in specifications to ensure that the references are appropriate. Cited standards should be incorporated properly,[7] and architects must not use standards if they do not intend to enforce them or if the standards are unenforceable, because "any provision of a specification which is not insisted on, only encourages violation of the intent of the contract" (RAIA AN14.01.100 1996g). RAIA AN05.01.600 (1996d) might also be regarded as a pro-standards document.

On the anti-standards side, the RAIA is concerned mostly about checking that the works comply with the specification, which it sees as potentially extremely onerous. It recommends that architects need not be familiar with materials and product standards, because the responsibility for conforming with these standards "rests with the manufacturer and/or supplier and their quality management systems" (RAIA AN05.01.001 1996b). As we will see, specifiers cannot cite standards correctly unless they are familiar with them.

Furthermore, a proposed change to the client-architect agreement would exclude from architectural service the checking of each item of work and materials supplied for compliance with referenced standards (RAIA AN10.01.110 1996e). One imagines principals insisting that no standards be cited at all, to ensure that the specification will be enforced. The Institute claims that this clause is only about the checking of each and every nail and screw, which of course is not a part of the architect's job, but the text still seems ambiguous.

An example of how the onus of responsibility for compliance with standards should be clarified is that the RAIA notes that the window manufacturer, although not a party to the contract, is responsible for the selection of aluminium extrusions (RAIA AN14.01.100 1996g). Yet the primary responsibility for compliance lies with the contractor, and this responsibility will be subcontracted on. The contract administrator inspects the works as appropriate to verify that the drawings and specification are being adhered to. And this will involve checking against product standards.

The worst thing about the RAIA advice is that it is internally inconsistent, being both pro- and anti-standards. This quote may help clarify things:

7. AN10.01.601 (1996f) advises using the title, number and date of the standard, and clearly stating which of any options within the standard apply. AN14.01.100 (1996g) suggests that the number of the standard is usually sufficient, and maybe the date. NATSPEC only gives the number.

- *Do not reference a standard you have not read.*
- *Use only current standards except in unusual circumstances.*
- *State any special project requirements that exceed the reference standards.*
- *Carefully review submittals to verify compliance of products and materials with the referenced standards* (Wyatt 1998).

Adhering to this sound advice is not easy. There are many standards.[8] They are relatively costly and are not always user-friendly or entirely relevant. Indeed they are often much less than perfect (contrasting with the general tenor of Standen [1995] which is that there are no deficiencies in standards).[9] They also date quickly. Foreign standards are particularly awkward to deal with. Accordingly, most of the issues mentioned are simply ignored in practice. This is not surprising given that to solve them is a daunting, if not impossible, task. To a large degree, the solution lies in the hands of the standards-producing agencies. Standards Australia is looking into these issues, for example, through its memorandum of understanding with the ABCB. Meanwhile, specifiers must do the best they can.

The first and most important point is to be crystal clear about which document is being referenced, in particular, which edition, incorporating what updates or amendments. The standards themselves do not at present identify editions and amendments in their own cross-references, but the BCA does. NATSPEC avoids the issue with a catch-all clause to the effect that editions current one month (or whatever) before close of tenders are to be used. The BCA's approach is to be preferred.

The general rule, to provide a defence against claims of negligence, is that the most current edition (plus amendments) of a standard must be used (Davie 1994). Superseded and withdrawn standards should not be cited. This only reveals how out of touch the specifier is. There are a few qualifications to this rule. First, it has been suggested (ACEA 1993) that under certain circumstances it may be prudent to comply with a draft standard if it sets more stringent standards than the current edition. Second, legislation might cite superseded standards or fail to cite the most recent amendments. To avoid breach-of-statute, one must comply with the referenced version rather than the current version.[10]

This leads to something of a dilemma for specifiers, perhaps best dealt with by working to (or above) the highest common denominator. Statutory standards generally only set minimum quality levels, after all. However, this presupposes that the specifier is familiar with the various editions and their amendments – a near impossible task, even in the relatively simple Australian situation.

A related thorny issue is that of secondary references in standards themselves. Because Standards Australia does not always keep up with its own publications, even on an annual basis, many standards become progressively more dangerous to use to the point where they may fall over because most or sometimes all of the referenced standards on which they rely have been superseded, or even withdrawn (AS CA55 [1970] is one example). AS 1884

8. Many complain there are too many building-related standards. As a result Standards Australia has a withdrawal policy and the ABCB also recommends standards to be deleted or consolidated (ABCB 1999). There are still almost 2000 building-related standards.

9. For instance, Standen (1995) says, "Given their immense value … standards deserve much more respect"; ".. the use of standards, unlike the standards themselves, is … partly good and partly bad".

10. In some state appendices of the BCA (Amendment 7), for example, a number of referenced standards have been superseded and a number of amendments are not referenced (Burke 2000).

(1985) still has not been updated to reference AS 3553 (1988), and continues to reference it as "AS XXXX". The ABCB (1999) had this to say on the subject:

When reviewing standards, confusion was caused by references when referenced standards had been amended, revised with a new number, or revised into a number of parts ... To overcome this confusion, the review [of over 2,000 standards] considered that the matter of standards referencing other standards or other documents needs improvement.

Use of dates and inclusion of amendments (as in the BCA) would help, for example.

Standards Australia has implemented a program of review that will take care of the worst offenders (i.e. standards older than 15 years). Nevertheless, the ABCB (1999) recommended that 385 standards be reviewed or updated, most being over a mere 10 years old. But secondary references in standards may still be out of date. For example, AS 3700 (1988), the masonry standard, had over half its cited standards superseded before it was republished. This is a serious problem, not least because these secondary references may be at odds with those cited directly in the project specification. It is difficult for users to know which standards apply and how far down the chain the courts will go.[11] At this stage perhaps all the specifier can do is be aware of the problem.

Standards can impinge on the conduct of a contract by spelling out the responsibilities of the principal or other parties. Fortunately this is rare in the more recent standards. Examples include AS 1884 clauses 1.3, 1.4.3 and 1.5 and AS 3610 (1995) Section 2 and Table 3.4.1. AS 3700 (1998) addresses the specifier in the text, stipulating documentation requirements (clause 1.4.1). Generally these are given in an "informative" Appendix, if at all (e.g. AS 4058 [1992] Appendix B) – such requirements do not belong in the body of the standard.

NATSPEC deals with this issue upfront, via the following subclause in *General requirements*:

Contractual relationships

Responsibilities and duties of the principal, contractor and superintendent are not altered by requirements in referenced documents.

Referenced standards are effectively extensions of the specification and so should follow the same principles in their drafting. For example specifications, like contracts, must be entirely mandatory, avoiding the use of 'should' or 'may'. And yet these terms are still widely used. Examples include AS/NZS 1677.1 (1998) clause 3.3.6 and AS 1884 Section 4. Conversely, AS/NZS 3017 (1996), for example, calls itself "guidelines" yet uses "must be", "is not", and "shall be".

Guide standards are another related issue. In the absence of standard specifications, guide standards are sometimes cited in specifications (and legislation) even though they are not intended for this purpose. Examples include AS/NZS 1768 (1991) – lightning protection, AS/NZS 2312 (1994) – protecting steel, AS 2818 (1993) – pool safety, and AS 3958 (various) – ceramic tiling. If called up in a specification (not a recommended

11. ABCB presumes that, in the BCA, only secondary references apply, and does not check tertiary and other references.

practice) such documents need to be 'hardened', perhaps with a clause along the lines of, "Comply with the recommendations of AS 3958."

However, this is not a very satisfactory solution. It is sometimes unclear whether the guide standard is recommending anything at all. It is better, but more tedious, to use the guide to help prepare a decent specification.

Standards might contain requirements or alternatives that conflict with legislative requirements or safe practice. Examples include AS 1604 (1997), which refers to pentachlorophenol at B5, AS/NZS 1677.1 (1998) Table 3.1, which references refrigerants that are either controlled ozone-depleting substances or transitional substances, and AS 3828 (1998), which clashes with general legal responsibilities for OH&S on building sites.

Some requirements are at odds with BCA96, including AS 4460 (1997), which cites a lighting standard superseded by that in the BCA. The ABCB (1999) review of over 2000 standards found five that were in conflict with BCA96. In such cases, the specifier is advised in the specification to override material at odds with BCA96 requirements. Perhaps one could rely, a little unfairly, on the contractor's obligation to comply with legislative requirements. Either way specifiers should be aware of the problem.

Other than standards, manufacturers' recommendations are the most commonly referenced documents within specifications. They are almost always cited in general terms i.e. the particular publication is not identified. This leads to problems because the material is continually upgraded. Different advice may well appear in glossy brochures, short and full technical literature, and in the information supplied with the product on-site. Which recommendations is the contractor supposed to conform to?

Problems also arise when different manufacturers of adjoining products give conflicting advice, as might happen with sealant manufacturers and structural glass manufacturers over preparation for sealing.

This book offers the same advice here as it did for standards: read it, make sure it is current, cite it properly, select options, and enforce it.

Coordination with other contracts

An increasing number of building projects involve more than one (prime) contract (as distinct from more than one subcontract) (in spite of Standen's [1995] suggestion that contractors "should not be hired to do building work"). These multiple prime contracts must be coordinated,[12] but by whom, and what is involved? There are essentially three such situations, as set out in Procurement at the beginning of this chapter:

- Single-prime contractor.
- Multiple-prime contractors.
- Nominated subcontractor.

Architects and engineers acting for the principal are usually responsible for coordinating

12. But see Standen (1995) on "builder's preliminaries".

contracts where there is a single-prime contract and other 'separate contracts' (see JCC Section 7) which may be on-site simultaneously, such as demolition or geotechnical surveying.

For fast-track or package contracting, in which the principal lets a series of multiple-prime contracts separately in a sequence to suit lead times and the construction program, a project or construction manager acting for the principal is usually responsible for contract coordination. Such coordination – dealing with 'who-does-what' – is essential in these circumstances so that all work required (including temporary work) is priced once and once only, and is properly integrated. This is not an easy task.

Finally, the relationship between a prime contract and an associated nominated subcontract must be clear. Because these are tendered at different times the architect and/or engineer, acting for the principal, needs to tell both the nominated subcontractor and contractor 'who-does-what'.

For these three circumstances it is legitimate, and indeed essential, for the contract documents to define 'who-does-what'. It is up to the principal to decide how the work is to be packaged, into separate prime contracts and/or nominated subcontracts. Likewise, the principal puts the work out to tender. Furthermore, as far as the individual contractor (A) is concerned, the principal will do any work outside the contract, engaging other contractors – B, C etc. – to do most of it. The contract documents for A cannot bind third parties such as B and C. Accordingly the principal (in reality, the principal's agent) is responsible for this coordination.

In passing, specifiers should be aware that the subcontract conditions themselves deal with coordination between contractor and subcontractor, particularly in terms of temporary services and facilities (see below). Do not duplicate this material elsewhere.

At the very least the contractor will need to know what work is included in the contract, which should be largely evident from the technical specification and drawings, and what is excluded. The contractor is especially interested in work boundaries or interfaces.

The contractor may also need to know who is responsible for the excluded work, to assist with on-site liaison. This advice can be given in the tendering material by providing the contractor with the documentation from relevant, separate contracts (see JCC clause 7.02.01), or by the project/construction manager during final procurement negotiations (or even on-site). This advice would be for information only – third parties cannot be bound by the contents of the contract. Very likely in real-life a mix of methods will be used. It is quite possible, for example, that precisely who will be doing excluded work will not be known when earlier package documentation is being prepared, or that documentation for other contracts may not yet exist.

Two main issues need to be covered: first, temporary services and facilities such as protection, scaffolding, power and cranage; and second, permanent works interfacing such as electrical cabling and connections for mechanical and hydraulic services, and where the work of the plumber (under the 'building' contractor, for example) stops and that of the mechanical contractor starts.

For multiple-prime contracts and for nominated subcontracts, provision of temporary services and facilities needs to be spelt out in the contract documents. For multiple-prime contracts a special interfacing section (NATSPEC's is called *Package definitions*) would be appropriate. For 'ordinary' subcontracts, apart from being the responsibility of the contractor, provision of temporary services and facilities is defined in the subcontract conditions (see SC-JCC clause 7.01 and Appendix Item E, and AS 4901 [1998] clause 4.4 and Annexure Part A), although this may need to be supplemented (perhaps as special conditions of subcontract). For nominated subcontracts also, the subcontract conditions are the appropriate place. These conditions are also a useful starting point for temporary services and facilities for multiple-prime contracts.

For multiple-prime contracts, permanent works interfacing with other contracts must be spelt out in the contract documents. Most of this could be done via the drawings and specification, which will describe in detail the work to be done in the contract, and in outline 'work by others' e.g. on the drawings a note saying "Lifting beam over, by others".

"Others" in fact equates to the principal, contractually. However, in terms of coordinating the documentation effort for the various contracts, a central schedule delineating 'who-does-what' (i.e. which contractor) would be very useful. Having prepared it for the use of documentors, it will be equally useful if passed on to the several separate contractors (at tendering) so minimising the risk of a common cause of dispute.

Such a schedule can have two levels: the contractual, defining the demarcation between the contractor and the principal, and the non-contractual, identifying the work to be done by other (third-party) contractors.

Contractually then, the schedule might define what work is included in the contract where this might be unclear, but more importantly it should define what work is excluded i.e. to be done by the principal. Non-contractually, just which other contractors are responsible for the excluded work can be stated here, but for information only. This document cannot bind them and cannot address them – they are outside the contract. If you must identify them, do so with care, ideally only in the tendering documentation.

For nominated subcontracts, commonly (but not always) tendered in advance of the contract(s) proper, interfacing must be clarified in the contract documents so that both the nominated subcontract tenderers, and the contract tenderers, know what their work boundaries are to be. Again a central schedule, much briefer than that for the separate contracts, can be very useful. The nominated subcontractor clearly needs this information at tendering but, because the contractor will assume responsibility for the work – and the coordination – of nominated subcontract(s), the contractor may have to redefine the boundaries or interfaces between the nominated subcontract and others. Accordingly the schedule should be provided for information only (clearly marked as not being part of the contract).[13] Otherwise, redefinition will require approval from the contract administrator, and may necessitate issue of variations.

The prime contractor is best informed about nominated subcontract boundaries via copies of the nominated subcontract documentation made available at tender, not via an 'interfaces schedule' in the contract documents.

Occasionally there are problems in this area, such as in a high-rise project where the building plumber stopped at the basement and the mechanical services plumber stopped at the roof, but the connecting pipework – 25-storeys worth – was missing from the documentation. No-one was responsible for it. It is hard to say how often this sort of problem occurs and also problems coordinating temporary works and services. However, it is fair to say that such problems are almost inevitable with the work of nominated subcontractors. This approach to procurement is still used, often for the wrong reasons e.g. procrastination, and sometimes even under contracts which do not deal with it, such as SBW-2. Generally, nominated subcontracts should be avoided as far as practicable.

'Ordinary' subcontractors may be under a single-prime contractor or under one of a series of multiple-prime contractors. Either way, the principal does not coordinate their work. By definition, subcontractors are not contracted to the principal, but to a contractor. The contractor resolves 'who-does-what' and the contractor puts the various 'ordinary' subcontract packages out to tender. Which subcontractor does what, and what is or is not provided by the contractor, is simply none of the specifier's business (approval of subcontractors may be, however, as in AS 4000 clause 9.2).

As we have seen, the various general conditions of contract imply that the contractor is responsible for their coordination. Specifiers must not interfere with this responsibility by, for example, addressing subcontractors directly in the specification or by providing 'scope of work' (and 'by others') statements for subcontract packages. Unfortunately, this is all too common, particularly for services specifiers.[14]

Contract documentation does not, and must not, differentiate the work of the various subcontractors. The floor plans, for example, are relevant to the work of perhaps scores of subcontractors and suppliers. The documentation cannot differentiate in any case, because the scope of the subcontract works is not known at the time of documentation (though assumptions are commonly made). However, contractors at tender will find a schedule of 'who-might-do-what' very useful, but such a schedule cannot be contractual. If provided at all, it must be for information only.

If this material is in the specification, then the contractor cannot depart from it without authorisation from the contract administrator, a fact all parties often overlook.

13. Australian standards use the term "informative" for this sort of material, and define this upfront.

14. Problems with "work by others" and other contractual issues of specifying are not confined to Australian services engineers. These topics and more were the subject of a seminar at the 1993 CSI convention in Houston (Shrive 1993). See also Gelder 1994e.

Equity

Contract documents identify and minimise risks, for the benefit of both parties to the contract. Risk must be allocated logically and fairly. Lazy specifying undermines this, to the detriment of both

- the contractor who, unable to accurately assess risk, has to decide the extent to which it will accept the risk or add it to the tender sum; and
- the principal, who will be faced with both higher-than-necessary and foolishly low tenders.[15]

If project requirements are not stated how can they be binding? In extreme cases the courts may overturn unreasonable requirements as inequitable. Design decisions should be proactive, carefully considered, and recorded in the contract documentation; they should not be reactive, ad hoc and made on-site by non-design personnel in a hurry. If it is important, spell it out in the specification. If it is not important e.g. you have no intention of enforcing it, say nothing. Be explicit, or be silent.

A common requirement is to comply 'with all relevant standards', or with a simple list of standards. This is inequitable on two counts: one, it implies that both the contractor and the contract administrator are familiar with all the standards and two, that these standards apply in their entirety, which of course they almost certainly do not (see next chapter).

Do not attempt to throw all risks and responsibility on the contractor. For example, do not use "... and do all other work that may be required to complete the building", or "The contractor shall furnish and include everything necessary for the full and complete construction of the building whether shown or specified or not shown or described". These 'murder clauses' are a sign of lazy or incompetent specifying. The specification could stop here! Specify what the other work is. These catch-all requirements may be covered in the general conditions of contract (not that this makes them any more acceptable), for example, PC-1 clause 8.2, for all work not mentioned, and JCC clause 6.01, only for obvious minor items of work not mentioned, which is not quite so drastic.[16] The requirements may also be covered by commonly accepted 'implied terms' (Bennett 1980, Chapter 6). *No Dispute* (NPWC/NBCC 1990) recommendation 1.2 states:

Obligations and/or risks within the control of the Principal should be borne by the Principal. Similarly obligations and/or risks within the control of the Contractor should be borne by the Contractor. There should be no discrepancy between the responsibility of a party regarding an obligation and/or risk and the party's authority to control or influence that risk.

This (a restatement of the Abrahamson principles mentioned elsewhere) impacts in a fundamental way on approvals and inspections by the contract administrator, leading directly to quality assurance, discussed below.

Risk allocation also impacts on specification terminology. For example, avoid the use of (common) words and phrases that are indeterminate, or subjective, such as

15. One indicator of bad (e.g. too much left unsaid) documentation is a wide range of tender prices.

16. Another example is provided in the UK contract JCT PWQ (1998a), clause 2.2.2, which states that "no ... inconsistency, error or omission shall vitiate this Contract", so an entire worksection or more could be omitted with impunity. The architect is expected to issue an instruction regarding the correction. Meanwhile, clause 8.1.2 has the contractor working to an appropriate standard of workmanship (but not materials) due to silence in the documents. They cannot both fill in the gaps!

- "... to the complete satisfaction of the contract administrator";
- "... as the contract administrator may direct";
- "... acceptable to the contract administrator";
- "... in the opinion of the contract administrator";
- "... a workmanlike job";
- "... a first-class job";
- "... sufficient"; or
- "... adequate".

These cannot be priced and the contractor cannot assess risk. They mean 'Guess what I will make you do'. Instead, specify objectively what is required. These 'guessing clauses' may make the contract administrator liable for future failure, so avoid them on those grounds also (again, an unfair allocation of risk). The specification must be fair to the principal as well as to the contractor.

Avoid the use of 'etc.' or 'and the like'. Be specific, so that a basis for pricing is clearly set.[17]

Approvals (of workmanship, of substitutions, of contractor's documents, etc.) carry risk for the contract administrator (and hence for the principal), much as 'guessing clauses' do. The RAIA's shop drawings stamp (see RAIA Advisory Note AN 16.01.100 [1995b]) recognises this. The DPWS (NSW) has advised specifiers to avoid the use of terms such as "to approval" and to specify hold points in the preliminaries only (i.e. not in the technical specification). NATSPEC defaults to witness points throughout its technical material.

In any case, the specification should not pre-empt the duties and decisions of the contract administrator. The roles of the contract administrator and principal should have been defined adequately in the general conditions.

Accordingly use:

- "The contract administrator may inspect the following: ..."; not
- "The contract administrator shall inspect the following: ...".

We have seen that some standards e.g. AS 3700 (clause 1.4.1) are problematic in this regard, committing the 'purchaser' to certain actions, such as the provision of information. The description of the works, embodied in the drawings and specification, is addressed solely to the contractor.

Adoption of Abrahamson risk allocation principles encourages the documentor and/or designer to take a hands-off approach on-site in lieu (largely) of traditional quality control. This recognises that the contract administrator is not in a position to exert control on-site[18] (see also Senior 1964). Accordingly, the specification is vital. Responsibility for execution and quality lies substantially with the contractor; the required quality must be clearly defined in the specification for this to work as the contractor cannot rely on the contract administrator for clarification and evaluation.

17. "It is better to be definite even if you are wrong; then, at least, there is a firm basis for negotiating the corrections" (Rosen and Heineman 1990).

18. However a 1996 court decision in Britain held that the architect is expected to superintend the works if the contractor is not competent to do so (Bingham 1996).

Part of this shift away from quality control by the contract administrator is the application of quality assurance to the construction process. Quality assurance, Abrahamson, and commonsense require that the specification be specific, explicit and quantifiable (or verifiable). Do not specify for excessive intervention and hand-holding by the contract administrator. *No Dispute* (NPWC/NBCC 1990) and quality assurance imply intervention should be minimised. It blurs responsibilities, and makes it impossible to quantify risk at tender. Trust the contractor you are using (e.g. consider partnering) or do not use that contractor at all. Vetting of tenderers is a vital part of this process.

Nevertheless, quality assurance alone does not totally remove the need for quality control by the contract administrator. Indeed, the specification must define the principal's quality control hold and witness points, to be incorporated in the contractor's inspection and test plan, and project audits invariably involve quality control. Anyway, for contractors as for manufacturers, certification to AS/NZS ISO 9002 (1994) is not proof of compliance with the product (building) specification and standards. Such compliance should be demonstrated separately, through quality control, the use of StandardsMarks, and so on (see Gelder 1997a).

In fact it is worth asking just what certification of a company to a quality assurance standard tells a consumer about that company's products and services. The answer is, not necessarily anything at all. For a start, quality assurance certification is about the company's processes, and not its products and services. The company, also, in writing procedures and setting performance criteria, will have selected a quality level that may be unacceptably low. Within reason, the certification auditor does not really care what quality level the company sets provided that it has been clearly defined in the quality manual and is followed through in the procedures and working practices.

Quality assurance certification of a company, however, is meaningful in two circumstances. First, if the company is a member of an organisation that insists on a well-publicised minimum quality level for its members, then quality assurance certification is a means of assuring customers that this quality level has been achieved.

Second, if the procurement process involves examination of quality manuals and the like, purchasers can then determine whether the selected quality level is acceptable. This may be the case in tendering for construction contracts, and in tendering for design work, if the principal requires it. The use of prequalification criteria may also take this into account, so saving the hassle of resubmitting the quality assurance manual at every tender.

Of course, those assessing the tender need to know what to look for. If the selected quality level is unacceptable then that tenderer could be rejected, or adjustments to the quality level could be negotiated. Alternatively the required quality level could be stipulated in the tender documentation.

For consumer goods and building products, claims from manufacturing companies that they have achieved quality assurance certification are pretty well meaningless to the consumer. Neither of the above circumstances apply. The company's selected quality level is unknown and unknowable – and may well be unacceptable.

Scope of work

This term is commonly misused in practice, for a wide range of concepts (Gelder 1994c). These concepts, and better terms for them, include the following:

- The Works.
- Outline description of the works.
- Package definitions (discussed above).
- Specification section content (discussed in Chapter 6, Contents and index).
- Schedules of work.

The prime function of the contract documents is to delineate "the Works". Indeed, the scope of work may be defined as everything described in the contract documents i.e. "the Works" (see JCC clause 1.02.04 and AS 4000 clause 1, for example). Since the contracts themselves use "the Works" (with a capital W) for this, it is to be preferred to 'scope of work'.

Outline descriptions of the contract as a whole are provided in the appendix to the general conditions of contract, and/or in the tender form. The inclusion of outlines (which often describe the work to be done, and the work not to be done) in the worksections or the *Preliminaries* is superfluous; it is time-consuming, and may conflict with the rest of the documentation e.g. by being incomplete, or by listing a component that is not required. It is little more than padding, and discourages estimators, among others, from reading the specification properly, particularly if misleadingly titled 'scope of work'. "Comprises the following" is an especially risky turn of phrase. It means that "the following" is a complete list. "Including the following" is a better phrase.

Last, schedules of work are commonly used in alterations and renovations projects, as the clearest way of communicating the extent of work required. They do not fully describe the scope of work, however, as they rely on other documentation, particularly the drawings and conventional specification material. In this respect they are similar to other multi-trades schedules such as those for doors and finishes but are often much more wide-ranging.

Summary

Perhaps the most critical point, contractually, is that the specifier must ensure that the specification is compatible with the procurement method. In particular, check who is responsible for interfacing, and who is responsible for design.

Coordinate the specification with the other contract documents. This means being familiar with them, and applies to both the lead specifier and co-specifiers. It includes using consistent terminology throughout all documents.

Resolve what will be specified, what will be scheduled, and what will be drawn, early in the process. Say it once, and in the right place: avoid repetition. This has many ramifications for specifiers and documentors at large. For example, the drawings and the specification must be exactly complementary with no gaps and no overlaps. In this respect, bills of quantities are problematic.

Be familiar with conditions of contract e.g. defaults for materials and workmanship, labour and materials, methods and coordination, giving notices and paying fees, and complying with the requirements of authorities. The specification may override the drawings, depending on the contract. Generally, conditions of contract require specifiers to specify workmanship in terms of outcomes (ends), not methods (means). Given the problems with bills of quantities, and with specifying methods, design-and-construct procurement may be used more in the future, as these problems do not arise.

Do not address entities outside the contract, such as subcontractors, tenderers and authorities. The contract administrator and principal should not be addressed in the specification, either.

Cross-reference explicitly (specification to drawings, and vice versa), if at all. Bills of quantities should cross-refer to drawings and specifications.

Be familiar with referenced documents, such as standards and manufacturer's recommendations. Be clear about the edition and amendments being cited. Beware of the content of referenced documents e.g. contractual conflicts, legislative conflict, problems with disclaimers and warranties, and status of guide text.

In the specification, deal with interfacing (permanent works, temporary works) between the prime contractor and other prime contractors, separate contractors, and nominated subcontractors, but not with 'ordinary' subcontractors.

Allocate risk logically and fairly and clearly, using Abrahamson risk allocation principles. Quantify requirements and avoid subjectivity e.g. "to approval". State requirements upfront. Do not rely on later actions to clarify what is wanted. Also state the method of verification.

Use other terms in preference to 'scope of work', depending on application e.g. the Works, outline description of the work, schedules of work, and package definitions.

6 Technical issues

There are two general[1] technical issues for the specifier:

- What are the appropriate quality levels for the project and its components?
- How are these best communicated to the contractor?

These questions are linked by the issue of 'choice' i.e. the degree to which the specification is closed or prescriptive. The more open the specification, the more the contractor is encouraged to shop around to find the best (cheapest complying) deal for both itself and the principal. Completely closed specifications generally imply that the specifier has already shopped around, and that what is specified is the best deal for the principal. Decisions on this topic need to be made for nearly every item specified, as well as for the contract as a whole.

This chapter looks at quality levels and at the agreement process for these levels. It briefly discusses the specification needs of design and construct, and approaches to briefing. The three main methods of specifying – performance, descriptive and proprietary – are then examined and compared against a background discussion of general principles. Other approaches to specifying currently in use are also reviewed.

Quality levels

Several factors drive the desired quality level of the project and its components. The main factor is, perhaps, anticipated life. Clearly it is false economy to build badly something that must last a long time, and vice versa. For buildings as a whole the anticipated life may range from a few weeks or months for temporary structures such as site sheds and exhibition spaces, to 200 years or more for monumental buildings such as the new Parliament House in Canberra, art galleries and cathedrals. The norm would be in the range of 20 to 50 years, related to financial considerations among others.

1. This is as close as the book gets to the nuts and bolts of specifying. If you want to know more about technical considerations in specifying a specific product, say, door hardware, refer to something like NATSPEC BASIC, including the Commentary, at least to start with.

A similar range also applies to components. Paint systems may be reapplied every year. Partition systems in commercial premises are stripped out and replaced for each tenant, perhaps every three years or so. Hotel suites are regularly refitted, driven by fashion in the marketplace. Building services in commercial buildings may be replaced every 15 to 20 years (e.g. airconditioning systems, driven by comfort and energy issues) or every 5 years (e.g. communication systems, driven by changing technologies). Prominent British architect Richard Rogers has argued that these services should be located externally as far as possible, to facilitate replacement with minimum disruption to the occupants (as in his Lloyd's of London building).

Building facades seem to have mixed life expectancies. They may outlive the rest of a 19th century building yet may be stripped regularly for a 20th century building to meet market expectations (as advised by real estate agents and property managers). Building structures tend to be the longest-lived components, particularly for buildings that must be disaster-proof, such as nuclear reactors and emergency services facilities.

Other factors that influence quality are often related to the building's anticipated life. These include the purpose of the building, required functional performance, appearance, and what Ferguson and Mitchell (1986) call "experience". The relative importance of quality and price will also vary with project type (see Table 6.1), and should be advised to designers (at briefing) and contractors (at tender).

Considerations for the purpose of the building include prestige versus utility, and flexibility versus permanence. Functional performance issues include design repetition versus one-offs, integration of function including multi-functionality of parts, specialist work or traditional work, and whether the philosophy behind the selection of fittings and finishes is green, high-tech, or conspicuous consumption.

The principal, the designer, the authorities and the local people may well not agree about the appearance of the building in which case different rules for quality may apply to different parts of the building. Even humble rooftop services may come under aesthetic control if a prominent neighbour looks out over them. Lastly, "experience" is about how

Type of project	Indicative quality:price ratio	
	For consultants	For contractors
Feasibility studies	80:20 to 90:10	Not applicable
Innovative projects	70:30 to 85:15	20:80 to 40:60
Complex projects	60:40 to 80:20	15:85 to 35:65
Straightforward projects	30:70 to 60:40	10:90 to 25:75
Repeat projects	10:90 to 30:70	5:95 to 10:90

Table 6.1: Quality:price ratios. The ratio is determined by the client and is used to weight award criteria at consultant or contractor selection (HM Treasury Procurement Group 1997).

the users perceive the building, with convenience, comfort, and ease of maintenance and repair all to be considered.

There is still more to the question of quality. Each component has attributes, of which some may be unimportant while others are vital. These attributes include appearance, strength, durability, hygiene and so on.[2] An in-ground structural or waterproofing component does not need to look good but it must have high strength and/or durability. Conversely the visual quality of a non-loadbearing ceiling in a reception area is important while the durability and strength are not – it can be fairly flimsy rather than robust. Some paint systems are applied principally for aesthetic reasons, and others principally for protection against corrosion. Their specifications will therefore be quite different.[3] Specifiers should come to each project, and each component, with an open mind.

In practice, specifiers do not formally deal with the issue of quality levels on a project-by-project basis. This would be very time-consuming, and in any case most buildings designed professionally fall into the same general range of a 20 to 50 year working life and, for a given practice, may be of the same building type. Assuming that appropriate decisions were made in the first place, they may not need to be revisited for every project.

However, mistakes can be made, but these will only come to light through post-occupancy evaluations or condition surveys. One such survey of a hospital (Sydgood 1991), for example, found that carpet and carpet tiles in corridors and common areas such as lounges, metal ceiling tiles (especially in wet areas), linoleum flooring, unprotected ward doors, vinyl strip skirtings, and sheet vinyl on corridor walls all failed to last the distance. Their durability had been overestimated. Unfortunately such evaluations-in-use are rare with designers apparently abandoning their projects at final completion (Bycroft 1989).

Similar decisions are sometimes made for all projects e.g. an architectural practice might have its own preferred basins, toilets, taps and so on, which are used, unquestioningly, on any residential, commercial, institutional and industrial project the practice may work on. Whether the quality of these items is appropriate to the project may not be considered. In fact among Bycroft's (1989) key deficiencies are "poor quality (domestic) fittings and furniture in high demand public amenities". The converse also happens, according to Burd (1994): "Under traditional contracts employers have almost certainly received a level of 'quality' that they have never recognised and, in many cases, do not need or want to pay for." The advantage of this approach (to the architect), of course, is that decision-making time is saved because, effectively, no decision is made!

The pursuit of quality can become obsessive. Burd suggests that, given the chance, most project architects will "pursue their concept of quality down to the consideration of every screwhead and panel pin, until exhaustion or bankruptcy intervenes". According to the Parliament House Construction Authority (PHCA) in Canberra, for example, the architect frequently produced late documentation, leading to construction delays and

2. Table 1 of AS 3854.2 (1990) – performance standards, provides a comprehensive list of attributes.

3. The New Zealand Building Regulations 1992 deals with durability in Approved Document B2: Durability (BIA 1998). For new products it can be very difficult to demonstrate compliance with the performance requirements of the lesser of the "specified intended life of the building" or

- 50 years for the structure, for difficult-to-access services and for hidden fixings and attached structures;

- 15 years for other fixings of the envelope and attached structures, the envelope itself, lining supports and other elements with moderate access but which are difficult to replace; and

- 5 years for linings, renewable protective coatings, fittings and other elements with ready access.

increased costs. A major factor "was the amount of time spent in reworking design issues".[4] Others have asserted that the cost increased because of the "unregulated pursuit of excellence", an assertion rejected by the PHCA because "such in-built inefficiency would have been self-regulatory because of the cost ... in unproductive salaries" (PHCA 1990). The PHCA has obviously never worked in a design practice![5] The lesson again is: know when to stop. This principle applies to the specification (and drawings) both quantitatively and qualitatively.

Quality now needs to be defined. Depending on criticality, the question of tolerances arises. That is, for certain components the contractor can be given quite a wide choice, for others the range must be narrowed and the specification therefore must be more precise. It is counterproductive to be precise where this is not appropriate, and vice versa. The principal will pay unnecessarily either way.

Quality needs for the project and its components must be communicated unambiguously at various stages of the design and documentation process, beginning with the principal's brief (Bycroft 1988). The principal's desires for quality and cost may be at odds; the principal may well want maximum quality for minimum price (see Figure 6.1). The designer will have a more realistic approach to cost, but may also wish to raise the quality level, in consideration of the designer's professionalism and design philosophy (the designer's hidden agenda), and of the perceived expectations of the community at large. Architects often seem to see themselves as the lone guardians of the community's right to award-winning buildings, in the face of philistine developers and authorities. The designer's modified view will be fed back to the principal for approval, and they may well compromise on an agreed quality level that will then be incorporated in the tender/contract documents.

The contractor has a different view, generally wishing to provide the cheapest complying solution, to get the job, and then to maximise profit (also see Figure 6.1), perhaps by substituting cheaper solutions. One function of the documentation is to limit the extent to which such hostile interpretations can be made. The contractor can, and must, build only what is literally specified or drawn, and should not read between the lines; this is particularly so under quality assurance. Accordingly ambiguity and errors in the specification and drawings, which create room for manoeuvre, are to the contractor's advantage.

4. A similar problem has surfaced for the Scottish parliament building. The project manager resigned, reportedly because of frustrations arising from the "fluid" designs of the architect designing the project, which it was claimed had the project running two months behind schedule and £20 million over budget (English 1999).

5. Architectural practices frequently operate at a loss, particularly on small projects (refer to RAIA Advisory Note AN02.03.701 [1995a]).

Design and construct

In the UK, as in Australia, there are three main approaches to design-and-construct procurement. The traditional approach, in which the contractor is involved from briefing onwards, comprised 20% of projects in 1995. Novation, in which the contractor is involved after briefing (commonly after sketch design) was the most common approach, accounting for 50% of projects. Develop and construct, handled in several different ways, ran to 20%.

These percentages are from the *University of Reading Design and Build Forum Report*

(Bennett et al 1996). The *Reading Report* is central to what is now the UK's Design Build Foundation, and its findings and recommendations remain influential. The Design-Build Institute of America (DBIA) also contributed to the *Report*. It suggests that the performance of design and construct depends "crucially" on two factors:

- The degree of detail to which the requirements of the principal and the design have been developed before the contractor is appointed:
  - Minimal (a statement of functions with little or no technical specification i.e. a brief).
  - Fairly detailed (plans, elevations, functional requirements with performance specification).
  - Very detailed (drawings with detailed prescriptive specification).
- The way in which the post-handover designer is selected:
  - Consultants chosen by contractor.
  - Novation of principal's consultants.
  - In-house consultants.

How this works is shown in Table 6.2, based on a survey carried out for the *Reading Report*. Minimal handover documentation (i.e. a brief) produces the best all-round result, and this necessitates the early involvement of the contractor. Full conventional

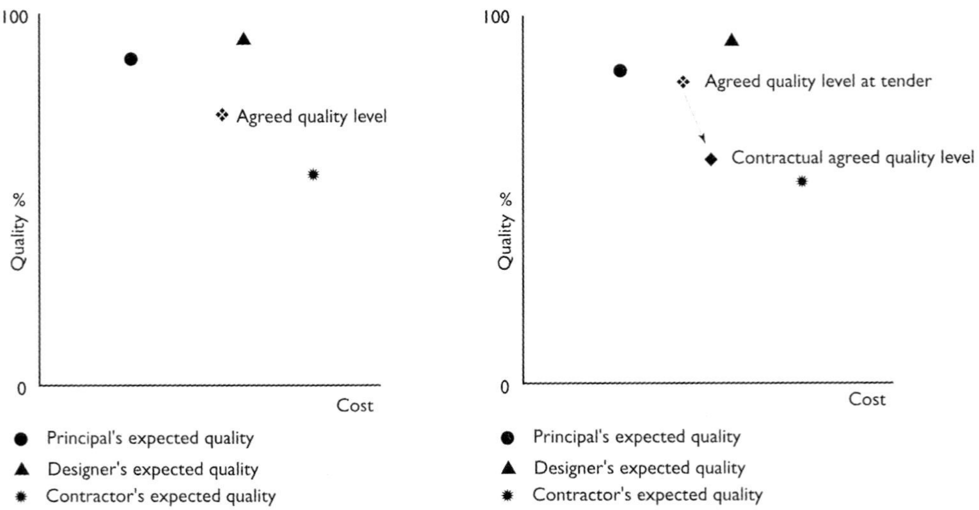

Figure 6.1: Agreed quality levels (based on Ferguson and Mitchell 1986). In principle, quality levels can be agreed in one shot under design and construct (left), with no subsequent shifts. In construct-only procurement (right) a shift in agreed quality levels is likely from documentation to execution. The more precise and objective the documents, the smaller the possible shift.

documentation gives average results overall, and half-way documentation e.g. sketch plans and performance specifications, gives the worst all-round outcome.

The key is continuity of design. This is best achieved, under design and construct, through the early involvement of a contractor (via a brief)[6] and use of in-house designers (no conflict of loyalties). Novation should not be used. The resulting conflict of loyalties is too disruptive to what looks on the face of it to be continuity of design. The *Report* does not recommend performance specification.

These conclusions might also apply to services specifications – either the installations should be full contractor design, or full consultant design. Half-and-half, which is current standard practice, does not work, as is noted below. Briefing for services design and construct, or the use of prescriptive services specifications for construct only, should be seriously considered.

Briefs

Briefs, the descriptive handover documents between the principal and the designers, are prepared for every project, either for consultant designers or a design-build contractor. And yet this critical phase in the procurement of a building, when mistakes can have major consequences, and where the effort is relatively small, does not generally have the benefit of any sophisticated standardised working tools.[7] It may be felt that such tools would constrain the briefing process, but given the feedback on the standard of briefing (critical omissions, too much detail and prescription, and so on), it is suggested that the benefits would outweigh the disadvantages, if any.

The brief is not, of course, organised around the worksection structure used in traditional specifications. It is not essentially about products – it is essentially about spaces (see Table 6.3). A suggested checklist for a brief, in outline, is given in Table 6.4 (from ISO 9699:1994). Reference should also be made to RAIA AN10.03.100 (1998b), which points out that the briefing process often overlaps with the design process.

Attribute	Completion on budget	Completion on time	Quality target met
Brief	Average: 32%	Best: 78%	Best: 68%
Performance specification	Worst: 22%	Average: 70%	Worst: 38%
Prescriptive specification	Best: 50%	Worst: 66%	Average: 48%
Design consultants chosen by contractor	Average for brief, otherwise no effect	No effect	Average: 53%
Design consultants novated to contractor	Worst for brief, otherwise no effect	No effect	Worst: 28%
Design consultants in-house	Best for brief, otherwise no effect	No effect	Best: 72%

Table 6.2: Getting the best from design and construct (Bennett et al 1996)

6. *Echoed in a US CII (Construction Industry Institute) study, which found that "success factors for D-B include* **less** *than 20 percent of the design is complete when engaging construction and excellent project team communication" [my bold] (Tibbs 1998).*

7. *Though tools do exist, such as CQR's Tessera, and the system developed in-house by the UK-based architect Foster Associates.*

This checklist prompts for facilities, construction entities, spaces, parts and products (both fabric and services). On the other hand the traditional specification, the subject of the rest of this book, only tackles products. Specification of the other entities can be in terms of performance, description or proprietary item, as for products. Even design methods can be specified in a brief (though it is usually unhelpful to specify design solutions, since this unduly constrains the designers). For example, designers could be required to use structural design standards like AS 3700 (1998), or access design standards like AS 1428 (various). These, it should be remembered, are not mandatory in the BCA, so their use is not guaranteed.

The *Reading Report* (Bennett et al 1996) notes that the lower the expectations for quality, the less likely they would be met under design and construct. This suggests that one should over-specify quality on low-quality projects, to increase the chances of achieving the quality levels actually required. Or principals should have high expectations, which stand a much better chance of being met (perhaps enforcement of the specification is an issue here). If the contract administrator (or equivalent) considers that quality is not important then little effort will be made to enforce requirements, and vice versa.

It is important in handover documentation at any stage in the procurement cycle that the designer states, and prices in, the approach to design verification or compliance. Where the handover documentation is a brief, then the design verification process for all later stages should probably be spelt out. This is quite distinct from the construction compliance process.

Entity	Status	Briefing		Design			Documentation			Construction
		Open brief	Closed brief	Developed brief	Sketch design	Developed design	Open documents	Compliance documents	Closed documents	Record documents
Spaces	Performance	100%	80% max.	20% min.	–	0%				
	Prescription	0%	20% min.	80% max.		100%				
Parts (includes systems)	Performance	0%	20% nom.	80% max.	–	20% min.	0%			
	Prescription	0%		20% min.		80% max.	100%			
Products	Performance	0%		5% nom.	–	20% nom.	95% max.	-	5% min.	0%
	Prescription	0%			5% exterior		5% min.	20% BCA	95% max.	100%

Table 6.3: Entities and their resolution in handover descriptive documentation

Value management is widely used in larger projects to ensure that the developing design is on track, representing a balanced solution to the principal's requirements as described in the brief. Generally the designer will talk through the developing solution with the principal to ensure compliance. A solid and detailed brief is essential if this process is to be meaningful. However, inexperienced principals can find this process difficult - it is not uncommon in building projects for the principal to view the finished building and exclaim that this is not what it had in mind at all. New virtual reality fly-through technologies, and simulations for other aspects of building design, should make this less likely to happen.

Public sector briefing systems have been developed for certain building types e.g. hospitals and schools. Competition briefs in the public domain are readily available to serve as models. Published master briefs are also available, such as the Institute of Housing /Royal Institute of British Architects' *Housing Design Brief* (IH/RIBA 1983), which classes its clauses into three levels of importance – essential, desirable and suggested – but which only addresses the design and performance part of the checklist in Table 6.4. Performance specifications for whole buildings (e.g. see Harrison and Keeble 1983) are available and these should fit into the definition of a brief used here.

Project identification
• Identity of the project • Purpose of the project • Scope of the project • Identity of the participants • Identity of other related groups
Context, aims and resources
• Project management • Laws, standards and codes • Financial and time constraints • Background and historical influences • Influence of site and surroundings • Client's future enterprise • Intended occupancy in detail • Intended effects of the project
Design and performance
• Site and surroundings • The building as a whole • Grouping of spaces • Spaces in detail • Building fabric performance • Plant, equipment and furnishings

Table 6.4: Brief structure

Methods – Overview

There are two broad classes of specification: open (contractor design), and closed or prescriptive (consultant design).

Under these are four basic methods for communicating the ends or outcomes required:[8]

- Performance (open).
- Descriptive (closed).
- Proprietary (also known as brand or outright).
- Construction methods (also closed).

Each can rely on custom-made text, or on reference to external documents such as standards. Reference is often regarded as a fourth method of specifying, but it is really a submethod (see Table 6.5). Each of these methods must be managed by the contract administrator, and the rules for management (substitutions, approvals, submissions, checking conformance) given in the specification.[9]

There are at least five desirable features of specification methods, as follows:

- Compatibility with the form of contract.
- Clear-cut responsibility for design.
- Internal consistency, both technically and contractually.
- Capacity to obtain a good price for the work.
- Ability to integrate and coordinate services, especially with structural, architectural and other services elements.

The first three are closely related.

Specifiers tend to specify certain items using the same method for all projects, regardless of the form of contract, except where the principal requires a different approach (e.g. public works authorities). Specifiers do not go back to first principles to work out the most appropriate method for the particular project (see Table 6.5 for examples). Just because a given item is normally specified this way does not mean that this is the only – or the best – way for that item to be specified in the particular project. Specifiers should have open minds on this topic. An appreciation of the issues is essential.

It is not the function of the documents to make a specifier's life easier. Rather, it is the function of the documents to serve *the principal's best interests* and if this means more work for the specifier, so be it. If fees are such that the specifier cannot resource this work, then that is a practice management problem. Specifying by proprietary item, and in particular allowing no substitutions, is a very attractive proposition if fees are tight. It may be justified on 'design' grounds, too. Specifiers and principals should be wary of this line of reasoning. It is spurious.

Design control *can* be maintained as far as it needs to be in more open specifications, with the backup of the drawings and through the approvals processes. An open

8. The contractor is generally responsible for the means or methods, though it is possible for workmanship to be specified either in terms of methods or outcomes.

9. For more on methods of specifying generally, see CSI (1996c).

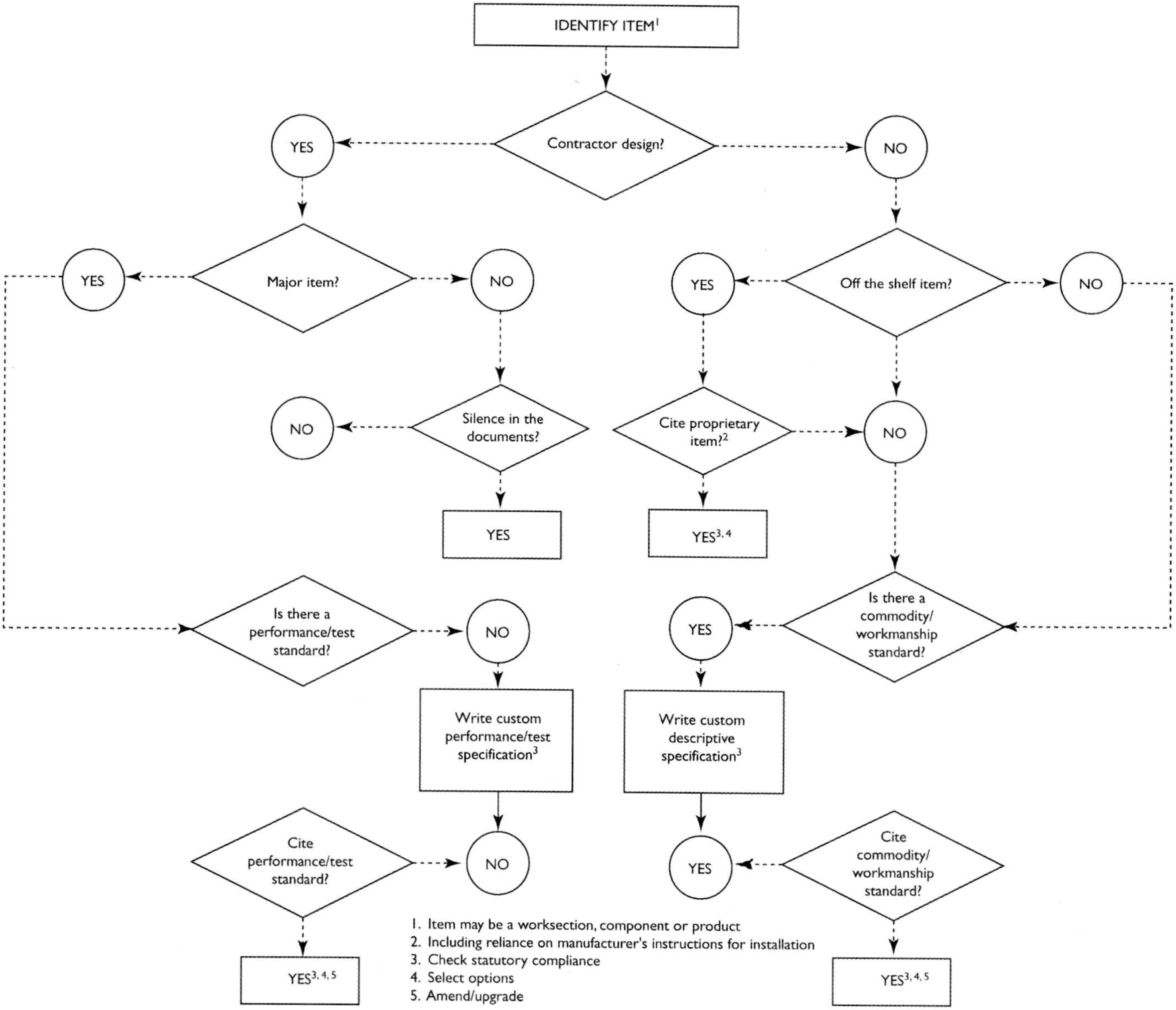

Figure 6.2: Specification methods decision tree

specification can still be tight as to the outcome (or at least, as tight as necessary). Try asking yourself just what it is about the proprietary item that compels you to call it up and consider specifying these attributes instead!

In many cases the contractor can propose substitutions (usually for elements specified by proprietary item) for what has been specified. Where approval is required, the contract administrator must be sure that what is proposed is at least as suitable as the item originally specified. It should be assumed by all parties that the original product (or performance, or standard) selection was not made lightly, that it is the fruit of considerable research by the specifier. Brennan (1993) states:

A contractor's discretion in substituting materials should be detailed in properly drafted plans and specifications. The contract must be examined carefully, rights against manufacturers considered, and the risk of assuming liability weighed against potential cost savings.

For a full cross-comparison, thorough submissions on proposed substitutions are necessary. These should be made during the tendering period, at least for major (identified) items. During construction, consideration of alternatives proposed by the contractor is likely to be hurried and ill-informed, and perhaps unduly influenced by the need to expedite the works, or to keep the contractor onside. Procedures for submissions should be spelt out clearly in the (non-contractual) tendering material and, if permitted during construction, in *Preliminaries* or *General requirements*. During tendering, approved substitutions should be advised to all tenderers.

Method	Submethod	Item
Performance	Custom	Curtain walls, airconditioning system
	Reference	Switchboard components, tests for curtain walls, safety of electrical equipment, access floors and tests, tests for soils, thermal insulation, efficiency of hydraulic fixtures
Descriptive	Custom	Groundworks, roadworks, terrazzo, stonework repair, landscape, parquet, rammed earth walling
	Reference	Painting, screws, demolition, adhesives, masonry walling, hydraulic systems, roof tile installation
Proprietary	Custom	Sanitaryware, door hardware, tiles, paints, carpets, plasterboard, bricks, plastic laminates, access floors
	Reference	Manufacturer's requirements
Construction methods	Custom	Floor sanding, concrete curing, roadworks equipment
	Reference	Codes of practice e.g. BS 8000 (various), manufacturer's instructions

Table 6.5: Typical specification methods used for selected items

Responsibility for the component's fitness for purpose varies with specification method and management system, as shown in Table 6.6 (which disregards installation). Responsibility for product selection is summarised in Figure 6.3. On silence in the documents, see Bingham (1995) who points out that where contractors fill gaps in specifications they are responsible for fitness for purpose.

The responsibility of the contractor for design decisions it may make was the subject of the Australian High Court case of Bryan v. Maloney (Barrymore 1995; Merritt 1995; Davie 1995b).

Performance specifications

Performance specifications are the most open type of specification. They are essentially a technical brief (i.e. product-focused), as they require the contractor to complete the design process. Their use, therefore, generally presupposes that the procurement route legitimately involves the contractor in design. Often simple selection of an off-the-shelf item against the criteria may well be all that is needed.

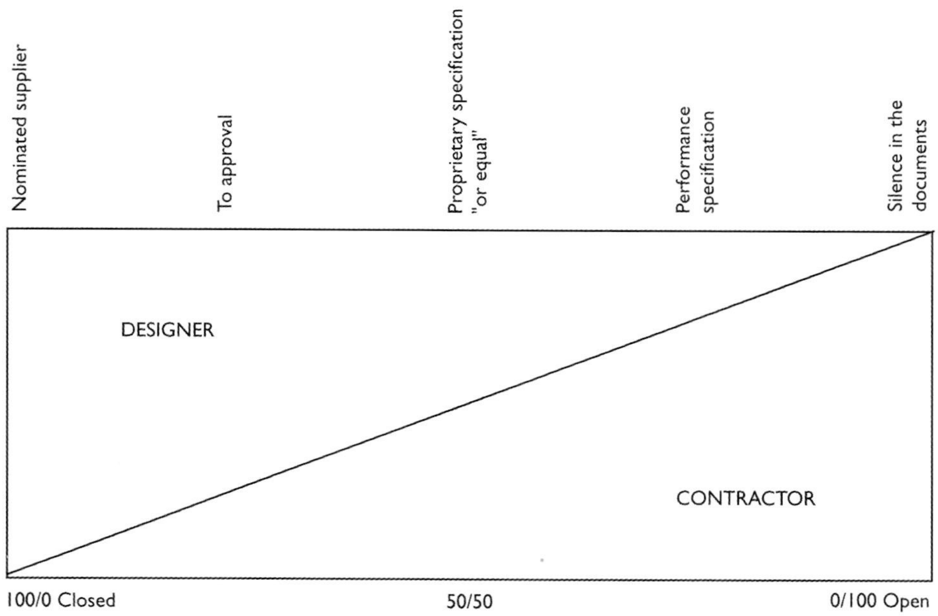

Figure 6.3: Spectrum of responsibility for product selection

Specification method and management system	Designer's responsibility	Contractor's responsibility	Notes
Open: performance specification	100% for performance i.e. performance, if met, will provide a fit-for-purpose solution (e.g. meets the brief). For construct-only procurement, the designer cannot properly delegate design responsibility to the contractor. The designer also remains 100% responsible for the solution	100% for solution (if design and construct) i.e. the solution meets the performance specification. The contractor has plenty of scope in devising a cost-effective complying solution	May use custom-made text, or may reference a standard (see below). The specification must state that the contractor must demonstrate (e.g. through type-test or project test reports) that the proposal meets the performance specification. If the specification does not state this, the contract administrator may reject it. There are no other grounds for rejection, so the performance specification needs to cover all important issues. Substitution, using different criteria or pass/fail points, is possible but rare
Open: silence in the documents –minor or routine items	0%	100%	A method of not specifying, but generally permitted under the contract. For example, see JCC (1994a) clause 6.01 Scope of the Works
Open: silence in the documents – major or new items	100%	0%	A method of not specifying. Designer is probably negligent in not specifying such items; an instruction filling the gap will be required. Clearly there is a grey area between the two extremes of minor/routine and major/new items, which varies from project to project. The specifier should err on the side of caution
Open: qualified silence in the documents, to approval e.g. 'Use an approved herbicide'	100% i.e. whatever is approved is fit for purpose	0% but the contractor must submit proposals for approval and has latitude in working out what may be acceptable	Another method of not specifying. Arbitrary – no grounds for approval are given, the contractor is unable to offer solutions with any confidence that they will be acceptable – it cannot be priced at tender, and there is a high risk of rejection. Approvals are hold points and will slow the job down. Implies that the specifier is aware that something should be said, but cannot think what it might be; decision is postponed
Prescriptive: descriptive specification	100%	0%	May use custom-made text, or may reference a standard (see below). This ratio applies only as far as the description itself goes – it may be incomplete. Substitution is possible but unlikely
Prescriptive: proprietary specification 'or equal'	100% if original proprietary item is used i.e. original item is warranted as fit for purpose	100% if a substitution is made i.e. substitution is fit for purpose. This gives a positive incentive for uncontrolled substitution, of cheaper goods, with no savings passed on to the client (the contractor will claim the substitutions were assumed at tender)	No approval is needed, so the contractor can substitute at its own risk. The specification should require notice of substitution, ideally at tender, at least for major (identified) items. The contract administrator nevertheless should be prepared to reject 'unequal' substitutions, so responsibilities are quite unclear. The basis for assessment is unclear (what attributes are important?) so the contractor will work to those attributes that suit the contractor, in spite of a fitness-for-purpose warranty (the contractor may not be fully aware of the purpose, or design intent, in any case)

Table 6.6: Specification method and management system

Specification method and management system	Designer's responsibility	Contractor's responsibility	Notes
Prescriptive: proprietary specification 'or equal approved'	100% i.e. the original or any approved substitute is fit for purpose	0%, but the contractor at least has room to move in terms of suggesting alternatives	Approval for substitution means that the contract administrator retains responsibility for selection. The specification should at least require notice of substitution, ideally at tender, at least for major (identified) items. Approval can be arduous and problematic. It must be timely and fair. The basis for assessment is unclear (what attributes are important?) and open to argument. There is a risk that the contract administrator's decision may be arbitrary
Prescriptive: proprietary specification, substitution is allowed if 'approved' ('or equal' is not used)	100% i.e. the original or any approved substitute is fit for purpose	0%, but the contractor at least has room to move in terms of suggesting alternatives	Approval for substitution means that the contract administrator retains responsibility for selection. The specification should require notice of substitution, ideally at tender, at least for major (identified) items. Approval can be arduous and problematic. It must be timely and fair. The basis for assessment is open. It is entirely up to the contract administrator and may be arbitrary; there is no appeal, and no certainty for the contractor. NATSPEC uses this approach
Prescriptive: proprietary specifications, no substitution allowed	100% i.e. implied guarantee that item is fit for purpose (and value for money). Specify that warranties are to be submitted, to at least cover fitness for purpose	0%	There is no evidence that the designer has shopped around. The supplier/installer can charge what it likes, and there is no incentive for competitive pricing. A bad result (economically) for the client is likely (unless there is a prior arrangement on costing with the principal, a major purchaser with leverage, for example). This can be embarrassing if the specified item is not available
Prescriptive: nominated supplier and/or subcontractor – a form of proprietary specification	100% i.e. implied guarantee that component supplied and/or installed will be fit for purpose. Warranties will usually have been obtained to offset this liability	0%	A firm quote has been obtained and built in as a prime cost sum or rate –this implies that there has been pre-tender negotiation i.e. the specifier has shopped around and found and agreed a good price. It may be that there has been a separate tendering process for this item
Reference specification: may be open (performance standard) or prescriptive (descriptive), and is often mixed	100% for appropriate and correct reference i.e. the standard is relevant, and defaults and options selected will lead to a solution that is fit for purpose	100% for solution's compliance with reference specification	The specification must state how the contractor is to demonstrate compliance with the reference specification. For some products and services this may be through third-party certification or, if this does not exist, through submission of test reports (tests may be in separate standards cited in the referenced standard). If it does not comply, the product or service may be rejected by the contract administrator. There are no other grounds for rejection, so the reference specification needs to cover all important issues. This method assumes that all parties are familiar with the reference specification. Substitution is possible for work overseas, where the contractor may be more comfortable with local standards

Table 6.6: continued

Theoretically, performance specifying encourages innovation, competition, and cost savings.[10] This is a very useful approach for areas of fast-moving technology such as electronics, telecommunications, and for one-off problems. It is also handy where the designers are out of their depth, and where the tenderers would be able to contribute relevant design expertise. A very good example is quoted in full in Kalms (1990) of US Signal Corps Specification No. 486 (1907) for a heavier-than-air flying machine. The Wright brothers were the successful tenderers.

Common applications of performance specification in building construction are fire-resistance, thermal insulation rating, and sound insulation (widely used in the BCA, and by specifiers ploughing through product literature looking at type-test results, but not much used in specifications, except perhaps for design-and-construct procurement). More common in specifications is the specification of concrete by strength (e.g. 20 MPa) rather than by mix.

The only way to evaluate compliance is to test the item (the test method, criteria and pass/fail points, and the testing agent, must also be specified), and such tests can be expensive, messy and time-consuming. Site or field tests may not give the same results as laboratory tests. On the other hand, where components have been type-tested (e.g. plasterboard partition systems), evidence that the performance criteria have been met can be readily demonstrated. Switchboards are a good example. For smaller classes, type-tested models are available whereas for larger classes, a project-specific prototype may have to be custom-tested to destruction. Curtain walling, being generally custom-designed for each building, is commonly custom-tested, using the method given in AS/NZS 4284 (1995), either at a testing facility or on-site, or both. Clearly custom-testing can be costly. Other examples include airconditioning systems (commissioning is, of course, custom-testing on-site), and concrete, for which the strength and slump (not the descriptive mix) are routinely specified and custom-tested.

The following artificial example of a kitchen benchtop matches those given elsewhere in this chapter, for comparison:[11]

Laminate – kitchen benchtop

Material: Decorative thermosetting laminated sheet.

Thickness: 1.2 mm ± 0.14 mm.

Colour: Match sample provided.

Finish: Match sample provided.

Fire performance: Test to AS/NZS 1530.3.[12]

- Spread of flame index: 0.
- Smoke developed index (maximum): 5.

Properties table: (see Table 6.7)

Clearly performance specifications can be quite wordy, unless a convenient test standard can be referenced, as for fire-resistance and the like (and indeed for plastic laminates).

10. For more on performance specifying, see CSI (1996b) and Gelder (1996b).

11. The contents are actually extracted from the standard AS 2924 (1987) – thermosetting laminates. This has since been replaced by AS/NZS 2924.1 (1998) but the new standard does not have individual test methods so, to make a point, we will pretend it does not exist.

12. Fire performance requirements are derived from the BCA.

These standards describe the test methods and criteria or attributes, but do not always give the pass/fail points. These must therefore be provided by the specifier.

Performance specifications could be combined with deemed-to-satisfy (but not mandatory) solutions, as in the BCA. This would expedite design. These pre-approved

Property	AS 2925 test method (since replaced by AS/NZS 2924.2-1998)	Property or attribute	Units maximum or minimum	Requirement
Appearance	2	Appearance		No defects visible from a viewing distance of 2 m
Resistance to surface wear	3	Abrasion to end-point	Revolutions, min	350
Resistance to immersion in boiling water	4	Increase in mass	% max	10
		Appearance	Grade (not worse than)	4
Resistance to dry heat	5	Appearance		No deterioration visible to the naked eye, other than loss of gloss
Dimensional stability	6A	Dimensional change	% max (length) % max (width)	0.45 0.90
		Appearance		No deterioration visible to the naked eye, other than loss of gloss
Resistance to cracking	8	Susceptibility	Grade (not worse than)	1
Resistance to staining	9	Appearance	Re-agents 1 to 18	No effect
			Re-agents 19 to 25	Moderate effect
Resistance to colour change in artificial light	10A	Carbon arc lamp	Min	4
Resistance to cigarette burns	11A	Appearance		No deterioration visible to the naked eye, other than loss of gloss and moderate brown staining

Table 6.7: Kitchen bench top laminate properties

solutions could be descriptive, proprietary or by reference to a standard, or any combination (including associated drawings). Use by the contractor of one of the deemed-to-satisfy solutions would save time and aggravation in the construction phase.[13] This would also meet the urge of specifiers to state *both* the problem and the solution (e.g. "To minimise corrosion use stainless steel fixings"), an unwise decision especially if the stated solution does not actually solve the problem!

Architects and engineers often specify by performance without even knowing they have done so. For example, they may ask for a "waterproof enclosure" to be provided to a piece of rooftop plant, or for "rigid supports adequate to sustain the load". To state the problem without a solution is to require design work of the contractor, which may not be appropriate. Likewise, to comply with a performance specification the contractor must do some design work. As stated in Chapter 5, standard forms of contract such as JCC (1994a) and AS 4000 (1997) are probably not up to major performance-specified (i.e. design and construct) activities. They assume the contractor is not responsible for design.

Strictly speaking, therefore, performance specifying (and therefore contractor design), at least for significant components, is not viable using the standard forms. The situation, including construct-only hybrids, is summarised in Table 6.8.

Nevertheless, architects and engineers often specify major items by performance as a matter of routine. On the building fabric side this approach might apply to GRP cladding or curtain walling. Use of this approach for building services is discussed more fully below.

An approach beloved of mechanical engineers in particular, is to use a manufacturer's proprietary catalogue as the source for performance data, and to specify by performance, on the assumption that similar products are available from several manufacturers, and will be acceptable. One reason this is done is that Trane chillers, for example, are only available from Trane. If specified outright, Trane could possibly charge what they like especially as they have no fixed catalogue price. Specifying by performance permits the contractor to shop around (seen as a process of selection rather than design), and encourages chiller

Procurement route	Contractor design
Construct only	Nil except perhaps for "minor items of work obviously to be done" (JCC 1994a and 1994b)
Construct only, with British-style amendment (i.e. JCT 98, PWQ98 Standard form of building contract, plus clause 42) (JCT PN 25 1993)	Partial, for certain non-critical items, based on performance specification
Construct only, with British-style amendment (i.e. JCT 98 plus Contractor's Design Portion Supplement) (JCT PN 25 1993)	Substantial, based on performance specification
Design and construct	Full, based on brief or performance specification (developed brief)

Table 6.8: Procurement and performance-based specifications

13. The CSI/DBIA (Construction Specifications Institute & Design-Build Institute of America) PerSpective specification system for design and construct does this. McClendon (1997), one of PerSpective's designers, calls this "inclusive multiple-choice". We have yet to see if it works!

manufacturers to offer competitive prices. That is the theory though practice is often somewhat different.

The source catalogue used by engineers often represents their preference. They would specify by proprietary item if they thought they could. Accordingly engineers, presented with a complying chiller from another manufacturer, will often say that this was not what they had in mind, thereby forcing the contractor to find out the name of the engineers' preferred manufacturer, and to use that after all. This is a farcical situation. A performance specification that is so tight that only one chiller on the market complies rather misses the point. Use of the performance method is supposed to maximise, not minimise, choice. The engineer has in effect specified by proprietary item, and the selected manufacturer/supplier can charge what it likes after all. And the contractor has no responsibility for product selection or fitness for purpose because no choice was possible.

It is essential when specifying by performance that detailed proposals, including a demonstration of probable compliance with performance requirements, be submitted at tender, at least for major (identified) items. This is because it takes time to evaluate a design and it is important to be confident that the tenderer can solve the problem.

Descriptive specifications

Descriptive specifications are the oldest type of specification – the early examples quoted in Appendix 1 (e.g. from Classical Greece) are of this type. The technique is still used, as in specifying colours, thicknesses, dimensions, and shapes. It is often the appropriate method for components assembled on-site. Such components cannot be purchased complete and so cannot be specified as proprietary items, or by reference to a manufacturing standard.

Decent drawings would replace most of a traditional descriptive specification these days. Even so, in the absence of other alternatives, this method (often supplemented by drawings) still makes good sense. The specifier may prepare custom material, reference a suitable descriptive standard, or do both to create a full description. Many standards include, or may even be predominantly, material of this sort. AS 3700 (1998), for example, describes the placement of wall ties in this manner.

Since the specification is commonly prepared late in documentation, the descriptive method is sometimes used to plug gaps in the details. The drawing staff may be flat out doing more important work, or there is no space on the drawings for a particular detail, so the specifier is brought in to create minor word pictures. However, this is not a good idea. Drawings are usually much clearer, and more concise, than their word-equivalents.

Descriptive specification usually gives the contractor leeway on assembly methods at least, although the components and workmanship may well be specified by reference to standards and/or to proprietary items. Compliance with the descriptive specification is checked by simple observation, as is compliance with the drawings, often made easier by comparing them with an approved sample installation on-site, built to define the qualitative issues.

Proprietary specifications

Proprietary specifications, in which brands and model numbers etc. are specified, can be very short, as this example shows:

Laminate – kitchen benchtop

Proprietary item: Laminex Premium Decorative Laminate.

Colour: Oyster Grey.

Finish: Texture.

This approach has been used since at least the late 19th century, as the examples quoted in Appendix 1 indicate (it has only really been possible since the industrial revolution and its factories replaced craft-made production). Brevity is perhaps one of its main attractions. Clarity is another. If done correctly, the specification is unlikely to be ambiguous. This approach also means that the specified item can be fully integrated into the design, a trite advantage for, say, doorknobs but significant for lifts and escalators. The downside is that substitution (see below) becomes very difficult.

Proprietary specifying is not used universally. A British survey, the *Barbour Report* (1994), suggests that architects may select up to "only" two-thirds of the products used in traditional procurement, and one-third in design-and-construct procurement.[14] For example, few specifiers would use it for premixed mortar, screws and bolts, and adhesives. Plus, it can only be used for commodities that can be purchased, or fabricated using a proprietary system. A different approach is needed for other items that must be fabricated (on- or off-site). In the US and Europe there is a move away from proprietary specifying, because it has a number of disadvantages, as outlined below.[15] Australian practice is similar, but there does not appear to be a move away from this method of specifying, which remains very popular among architects.

Proprietary specifications are perhaps the most closed of the four types of specification. The specifier generally is entirely responsible for selection. The contractor has little leeway, other than to shop around for the best prices. The item is usually available from a variety of sources and can usually be installed by a number of organisations.

But this is not always so. The desired product (e.g. an imported roofing system) may be available from only one supplier, or be installed by only one licensed installer. The supplier or installer can charge whatever it likes for the product or service, so the best result for the principal may therefore not be obtained.

Specifying by proprietary items is perhaps one facet of architects and other designers resorting "to conservative design solutions", which substantially deny clients both "cost savings in construction [and] innovative design solutions" (Peck 1995).

There are several ways around the problem of closure.

One way is for the specifier to shop around in advance and tie the chosen (or only available) supplier/installer to a fixed price, in which case it should become a nominated supplier/subcontractor, with the price built into the contract as a prime cost sum. To nominate

14. The Barbour Report 1994, published by Barbour Index, cited in Billingham (1994).

15. For more on the move away from proprietary specifying, see Kasimer and Miller (1992), Rosen and Heineman (1990) and Rosen (1999). For more on American practices, see Olmstead (1996).

the supplier/subcontractor without also locking them into an agreed price is asking for trouble.

Nomination must include the agreed contractual sum for the item. The contractor will subsume the nominated supplier/subcontractor, pay them this sum (the prime cost sum), and claim both the prime cost sum and separate overheads (for coordination, storage, handling and so on) and profit. This approach makes good sense for fabricated items, where the specifier is shopping around both for price and quality of fabrication (e.g. for a custom-made stainless steel sink where not all stainless steel fabricators will be able to produce the required qualities). But for most off-the-shelf items, where the quality of the product will not vary from supplier to supplier and the contractor can be trusted to get the best price, simple specification by proprietary item is enough. To assess the contractor's claim for the work, the contract administrator will need to have an idea of what the prime cost will be, though this is not written in the contract (see also Unknowns).

Another way is to identify in the specification more than one equivalent product, or a list of selected suppliers/subcontractors. This is not a bad idea as it allows the contractor to shop around for the product with the best price. Another option is for the principal to do the shopping around. For example, when public works authorities require certain lock systems to be used a favourable supply deal is arranged so the manufacturers will not be able to charge what they please when their locks are specified outright. This also applies for principal-supplied items (supplied at no cost to the contractor), for which the contractor will be able to claim overheads and profit.

In certain circumstances, substitutions are possible at the behest of the contractor. This may be on the grounds of cost (i.e. a cheaper equivalent is available), or availability, but performance must not be compromised. To guard against compromise, a complete specification is needed. Substitution can lead to bad feeling among trade representatives who believe they have 'sold' a product, in that it has been incorporated in the specification, only to find another product being used on-site. They may not be inclined to give such good service next time around.

Substitution is problematic in other ways. For example, in Britain contractors (maybe 40%) and subcontractors (25%) are making substitutions without the consent of the principal (i.e. the architect) who remains responsible for performance (Rogers 1993).[16] In Australia, substitutions (by the architect) have also been known to repudiate the contract (RAIA Advisory Note AN15.01.402 [1986]).

On managing substitution, do not use "or equal". This implies that substitutions are encouraged, and that the question of just what is equal is only the contractor's to resolve. This is a recipe for disaster. Nielson and Nielson (1981) advise that "or equal" specifications are the primary cause of around 25% of disputes involving specifications. "Or equal approved" (intended to imply that the contract administrator has the final say, which rather slows the job down) is not much of an improvement.[17]

16. For more on the practice of substitution in Britain, see Billingham (1994) which discusses substitutions at length, from the points of view of the contractor, subcontractor, manufacturer and specifier.

Brennan (1993) suggests generally that a "... contractor proceeds at his own peril ... when implementing products or methods contrary to the plans or specifications. Contractors assume the risk of any defect or extra expense associated with an unapproved selection".

17. Hartog (1992) mentions a North American court case in which 15 months passed while the contractor and designers disputed approval of a hospital bedlight fixture submitted as equal to the brand name specified. A court found the substitute "unquestionably equal" and held the architect liable for the cost of construction delays.

"Or equal" is rarely to be taken literally in any case. Usually only a few characteristics are critical and the rest are negotiable or unimportant. Just which are critical is rarely stated or obvious, and so is in danger of becoming a subjective matter for the contract administrator to resolve. This often leads to disputes. The risk to the contractor is not priceable at tender, either. Nevertheless, both terms unfortunately remain widely used.

The Northern Territory Department of Transport and Works follows proprietary specifications with a list of criteria. This is done to avoid arguments about which characteristics are critical but runs the risk of conflicting with the identified item's characteristics.[18] Some specifiers expressly forbid any substitution, which is safe if you are sure that all proprietary items are available, in a timely manner, otherwise this approach could backfire.[19] NATSPEC *General requirements (Proprietary items)* permits substitution where specification is by proprietary item, provided certain conditions are met. These include approval. The particular phrasing (op cit.) is intended to stop ambit substitutions by the contractor, made with a view to maximising its profit.[20]

Contractors sometimes advise that the specified product is not available simply because they do not know where to buy it. Rather than appearing foolish they will take a chance that the specifier has not done adequate research, and does not know where to buy it either. This ruse succeeds occasionally but can be defeated if the specifier has checked availability and sources, and/or if at least one supplier is identified "for information only" i.e. not nominated. Including contact information for the source(s) also helps, provided it is clear that a particular supplier is not being nominated. This approach is especially useful for unusual or imported items that may not be available locally.

A commonly stated disadvantage of the proprietary approach is that it can be unfair, anti-competitive or just plain biased. There have been court cases in the US in which non-specified manufacturers have sued the principal.[21] One threatened case in Britain revolved around the BSI not having quality assured painting contractors on the pre-qualification list for its new building, even though BSI wrote and promulgated the quality assurance standards! Overall, a quarter of the subcontractors were not certified (Ostler 1994).

In anticipation of this problem most public works authorities around Australia have long insisted that specifications be open, to encourage competition in the marketplace (and to minimise the number of complaints ministers receive from disgruntled manufacturers). Nevertheless, public works authority attitudes to specifying by proprietary item vary considerably. Tasmania, for instance, allows free use of proprietary specifying; the Department of Urban Services (Australian Capital Territory) allows the proprietary approach only for Australian-made items; "similar and equal to" is not allowed by the DPWS (NSW).

In the US, procurement practices for public work (i.e. work funded by taxes) must, by law, allow all taxpayers an equal opportunity to compete for that work (Dorsett 1993). Accordingly proprietary specifications are not used. Similar principles for 'public' works

*18. Wolfe (1994) says, "In any case, specifications must include all features and performance criteria deemed necessary for a product, **even if a brand name is used**" (his bold).*

19. Consideration of life-cycle costs, particularly repairs, may justify a 'no substitutions' policy for some items, since some products may have better reliability than others, or spares and training can be rationalised if new equipment matches old. Running costs and salvage value should also be considered (Wolfe 1994).

20. Hutchinson (1993) suggests that "... the contractor can use considerable pressure, particularly in the latter stages of a contract, to suggest cheaper alternatives to save money ... to absorb early contract extras and, of course, at some gain to himself".

21. Dorsett (1993) quotes a manufacturer's representative as saying, "Architects have every right to limit their choices and by doing so do not break any laws", and that "Architects should not be intimidated into accepting a substitution by a manufacturer threatening a lawsuit".

(broadly defined) apply to specifying in Europe (e.g. the UK), the idea being that all relevant European manufacturers must be able to compete for the work. There is also the general public sector obligation to be seen to be providing the taxpayer with value for money. This generally means not using proprietary specifications as these are open to the accusation that selection was not competitive, or even that it was corrupt. Compulsory competitive tendering, even for the smallest item, has been the trend in this sector.

Architects commonly use proprietary specification for items where aesthetics are the overriding factor. This is not always a proper use of this method, as it reflects the architect's hidden agenda as designer. That is, precision is rated as necessary for these items, but not for others. Other methods may be more beneficial to the principal for aesthetic items, and proprietary specifying may be the right way to go for non-aesthetic items. Selection, too, may not be on the basis of careful project-specific research but on the basis of routine: "we always specify this handbasin in commercial work".

From the principal's point of view other factors and other components (e.g. wall ties) may be more critical, better justifying the use of the closed proprietary method of specifying with its onerous liability. The hard-bitten and hard-pressed contract administrator may well also take a different view of the relevant attributes under an "or equal" approved regime, perhaps more with a view to expedience on-site than to aesthetics.

Proprietary specification is easily followed up on-site. Look for labels on the objects themselves, or on the packaging, and check documentation such as receipts. Ideally, do all three. In the supply of natural commodities such as stone and sand, it is not enough to cite the (proprietary) source and/or supplier in the specification as quality and characteristics will vary considerably. Required attributes (e.g. colour, grain, grade) must also be stated along with requirements for batching, mixing, culling, sampling and the like. Follow-up on-site is inevitably more complex, involving careful inspection. In extreme circumstances it may even involve tracking the commodity (e.g. a timber veneer) from its source to the site (Gelder 1994f; Willis and Tonkin 1998).

Reference specifications

With reference specification, external documents such as (most commonly) an Australian standard specification are cited, or documents such as APAS's GPC specifications, or even a BCA specification.[22] The method is generally not possible for new or unusual items. Development of standards usually lags behind development of products, and interim solutions such as CSIRO Building Products and Systems Appraisals (usually known as CSIRO Appraisals and replacing ABSAC Technical Opinions) or overseas standards or assessments may help.

The method has only been possible in Australia since the early 20th century (Standards Australia did not exist pre-1920s). Reliance on manufacturer's written instructions can often reduce reliance on standards e.g. for erection of plasterboard, for painting. Pre-

22. RAIA Advisory Notes AN10.01.601 (1996f) and AN14.01.100 (1996g) (Gelder 1993a).

editing and customising of master specifications, which commonly contain many references to standards, will also reduce reliance on standards, particularly as proprietary items are substituted for open specifications making the specification more closed. But is a reduction in reliance on standards necessarily a good thing?

Specifiers sometimes try to avoid using standards by

- saying little or nothing on the subject, which is not usually appropriate for the sort of projects dealt with by professional specifiers;
- writing their own material, which is time-consuming, may make compliance uneconomical or non-viable, and makes the specification very bulky; or
- quoting or paraphrasing standards, often out-of-context. This is error-prone, again producing unnecessarily bulky specifications, and introducing issues of copyright and currency.

It is clearly better to reference accepted current standards where appropriate e.g. for standard components and materials, rather than attempting any of the above. Not using accepted standards might mean that

- BCA performance requirements are not met;
- unacceptably low-grade imports may be used (e.g. galvanizing for fasteners);
- insurance requirements are not met (e.g. security screen doors);
- supplier confusion arises;
- extra costs are incurred; and
- the efficacy of the standards system is undermined.

The following is equivalent to the proprietary example given above. One immediately obvious disadvantage is that it is slightly longer, another – the major problem – is that it relies on the tenderers and contractor (and specifier) knowing the contents of the standard(s):

Laminate – kitchen benchtop

Material: Decorative thermosetting laminated sheet.

Standard: To AS 2924, Class HGS.

Thickness: 1.2 mm.

Colour: To AS 2700, N41.

Finish: Textured.

Spread of flame index to AS/NZS 1530.3: 0.

Smoke developed index to AS/NZS 1530.3 (maximum): 5.

StandardsMark: Required.[23]

Standards are available for manufactured products (both performance-based and descriptive in some cases), on-site workmanship, and methods of test (geared to performance-based specifying). This approach is fairly open. The contractor must shop

23. StandardsMark was an optional requirement of this standard, which is now superseded.

around to find products or whatever which comply, and is commonly required to submit selections for approval.

To give the contract administrator more time to determine compliance, it is suggested that proposals for critical (identified) products specified by reference be submitted at tender.

One problem with compliance is that manufacturers may claim their products comply with standards but then have difficulty demonstrating it. Most manufacturing standards contain or reference test methods and criteria. Have these tests been successfully carried out? Sometimes they have and sometimes they have not, but in both cases specifiers should be given the test results to work this out for themselves (effectively carrying out second-party certification, rather than simply relying on first-party certification).[24]

The simplest solution to the compliance issue is to use third-party certification such as the StandardsMark for Australian standards, or inclusion on the product registers of organisations such as the SSL and APAS.[25]

Requiring such certification from manufacturers can rather limit the field of potential suppliers. In extreme cases such a requirement effectively closes the specification right down to become the specification of a single proprietary item. This affects outcome in terms of liability (the specifier's is increased) and in terms of getting the best deal for the principal. It also defeats the purpose of using this method. So certification is very useful, but is not without hazards.

One overall limitation of product certification is that it is largely geared to satisfying statutory performance requirements only. However, certification schemes of general application to specifying are becoming more numerous (see Chapter 2). All that's needed now is for specifiers to use them!

The sheer quantity of relevant standards creates enormous problems for specifiers and other users. Many are used in legislation and in master specifications systems such as NATSPEC.[26] The cost of acquiring and updating standards are major related problems (in comparison, manufacturers' literature is generally free).

These standards in turn cite others – for example, AS 3700 (1998) references 49 (including standards for masonry units) – so the number of relevant standards rapidly multiplies. Strictly, specifiers are expected to be familiar with the standards they cite (Davie 1994). They should certainly be familiar with those cited in the BCA.[27]

The issue of familiarity is a major stumbling block in the proper use of standards. *No Dispute* (NPWC/NBCC 1990) stated that:

MB-CHAA [now MBA Inc.] claim that a selective reference approach results in a contractor having to refer to a considerable number of standards to obtain the full description of the work to be performed and is therefore particularly onerous in small to medium size projects, where it cannot be expected that all the standards listed in a specification can be kept on site. They oppose the use of the selective reference form of NATSPEC.

This situation regarding standards seems onerous to many practices. But be thankful this

24. In some cases, the answer may be that nobody knows. In which case the chances are that they have not been carried out, and should be.

25. Surprisingly, a survey carried out by NATSPEC suggests that few specifiers know or care about third-party certification, in spite of its long history (see Gelder 1993c, 1994d and 1997a).

26. 'Selective reference' is normal international practice, for open master specification systems, including the NBS National Building Specification (Britain), MASTERSPEC (US), and the STABU System (The Netherlands).

27. RAIA Advisory Note AN14.01.100 (1996g) and the CSI Manual of practice: Construction documents fundamentals and formats module (1996a) also recommend familiarity. To argue unfamiliarity with standards as a reason to avoid specifying by reference is in many instances an abrogation of professional responsibility.

is not the United States, which has over 90 standards-producing organisations (Gelder 1994a; Bullen 1991).

In spite of all these problems, specifiers cannot avoid the use of standards. Fasteners, premixed mortars, adhesives and the like are most efficiently specified in this way, for example. Further, compliance with statutory requirements is often demonstrated most readily through the contractual requirement to comply with a standard. For masonry, for example, the usual method of specifying is to cite AS 3700 (1998) upfront. This is one of many deemed-to-satisfy – not mandatory – standards in the BCA.

The specification must spell out how the contractor is to meet optional BCA and other statutory requirements otherwise the contractor will have to work this out, which would be a grossly unfair allocation of risk, for traditional construct-only contracts in particular. This statutory conformance use of standards in specifications will become more important now that the BCA is entirely performance-based (see Chapter 2).

Given that the use of at least some standards in specifications is unavoidable, specifiers must understand how to cite them properly. Familiarity is clearly essential, not least because of the standards' limitations and imperfections, as mentioned in the previous chapter (e.g. quasi-contractual text). The main point on citation is that standards should be referenced meaningfully and helpfully and so that the contract administrator can readily check compliance.

Blanket reference to all (or to all or any, relevant) standards does not meet these criteria (Hall 1994), nor does blanket reference to a list of selected standards.[28] Blanket reference even to a single standard may be hazardous. Some have so many options built into them that blanket citation is pretty well meaningless. A good (or bad) example is AS 2688 (1984) – joinery doors. Among other things, it describes eight types of door, offers five grades of timber, four classes of construction, and provides for panels of plywood, hardboard, glass or particleboard.

More standards are providing for several grades or levels of quality. The specification must state the grade required. Some standards pull all the choices together in a summary, rather than just leaving them buried in the text. Certain standards provide assistance in making the choices. But most standards offer no assistance at all on the selection of grades. And very few are self-selecting.

Standards are best referenced at the appropriate point in the specification, and in such a way that their intended application (if restricted) is clear. For example, AS 3700 covers the design of masonry, materials and construction. In projects where the contractor is not expected to design the masonry, blanket reference is misleading. Is the contractor expected to redesign the masonry, and who then takes responsibility for the design?

If the design has been done, the specification would say "Masonry materials, construction and detailing: To AS 3700." Otherwise it would just say "Masonry: To

28. A worst-case scenario is the blanket reference of around 400 standards for curtain walling, including from Australia, Singapore, Japan and the US, in a curtain walling specification for a project in Thailand (Hartog 1994b).

AS 3700". Standards Australia is helping by producing separate standards and handbooks for maintenance and for design.

Standen (1995) advocates identifying relevant clauses in the citations e.g. "To AS 1234 clauses 4.1.1 and 4.1.2". Apart from being cumbersome, clauses should not be cited in isolation as they invariably rely on other clauses in the standard for correct interpretation. It is better to limit the citation by the key word used, e.g. "Pavement bars: To AS 1742.2" will automatically home in on clause 3.3, but will also pick up any other related material in the standard.

Usually variations to the standard e.g. overriding unwanted defaults, and selection of options where the standard offers choices (as many do) should immediately follow citation.

Specifiers commonly rely on a blanket instruction to comply with "applicable portions" of a list of standards. This is usually either inadequate – these portions contain options, gaps and defaults the specifier may not like – or impossible. It is also extremely onerous for all concerned with the construction phase, if taken literally (as it should be, and as the courts certainly will in a dispute). A list of standards does not comprise a specification.

Likewise, in most specifications reliance on manufacturer's instructions is generic, with particular editions, or rules for working this out, not being identified. And yet these documents are part of the specification, just as referenced standards are, and should be treated in the same manner – even more strictly, really, since they often mix promotional material with technical advice and specific instructions. Beware of inconsistencies with cited standards, the specification itself, and good specification practice. Methods are often specified, for example. Warranties and guarantees must also be properly dealt with.

Standards have their limitations, and specifiers need to evaluate the scope and currency of the standard first.[29] On scope, some standards are domestic, not commercial, others are restricted to electrical installations less than 1000 V, and so on. The above-mentioned AS 2688 is limited in scope to certain types and sizes of timber doors. It does not apply to fire doors, hospital doors and larger-than-normal doors, is silent on decorative mouldings, and does not deal with door hardware or door frames. Likewise, AS 2896 (1998) – medical gas systems, does not deal with suction systems for laboratories or hospital dental units, or the supply of oxygen from pressure swing absorption.

Not all standards are current, entirely relevant or widely accepted. Standards are sometimes even available before the product, as in the locks standard, AS 4145.2 (1993). The lesson is that specifiers and others also need to understand the applicability of the standards they use. Even the BCA deemed-to-satisfy AS 3700 is not well understood by bricklaying contractors (Nawar and Zourtos 1994).

Defaults in standards (i.e. requirements that apply unless overridden, which is only to be done advisedly) may not actually be what the specifier has in mind. Many standards anticipate project-specific variation of defaults. For example, in AS 1214 (1983) the

29. Standards Australia has a withdrawal policy. Under this system, standards over 15 years old will be automatically withdrawn unless there is a reason for retaining the standard.

method of determining the coating mass in case of dispute has been selected, but specifiers may nominate an alternative method. In AS 1897 the thickness of zinc coatings is set at 5 μm "unless otherwise specified". Tables in the standard suggest other thicknesses for the less common applications.

Performance specifying using standards is often straightforward (as in the laminate example used here), but for some products the standard says surprisingly little. For instance, not all relevant properties may be dealt with, test methods may not be mandated or even provided, pass/fail criteria[30] may not be given. A good (or bad) example is AS/NZS 4455 (masonry units), which covers dimensional deviation satisfactorily, but does not quantify any other identified properties. These include unconfined compressive strength, breaking load, integrity, coefficients of contraction and expansion, density, lateral modulus of rupture, abrasion and slip-resistance, resistance to salt attack, pitting due to lime particles, resistance to freeze/thaw, and efflorescence. A decent performance specification for clay bricks would barely make any reference to this standard. Its predecessor for clay bricks (AS 1225 [1984]) provided much better, if still incomplete, coverage of these properties.

Some performance standards provide deemed-to-satisfy solutions for those preferring a prescriptive approach to specifying e.g. AS 2050 (1995).

Australian standards, being consensus documents, do not always set quality levels that will suit principal or designer requirements. Accordingly, a number of organisations representing purchasers have established their own standards. APAS is a good example. Paint manufacturers use GPC specifications (produced by APAS) in preference to the Australian (and New Zealand) standards; they are cited on cans and in catalogues, whereas the standards usually are not. Nevertheless, relevant Australian standards are used as the basis for GPC specifications, which are not used only on government projects.

GPC specifications set a good standard, but better ones are available, as in the premium versus trade ranges of various manufacturers. Specifiers should check the manufacturer range.

Construction methods

Building procurement has three strands: definition of the problem (the principal's task), design of the solution (the designer's task), and construction of the solution (the contractor's task).

Just as it is inappropriate for the principal to pre-empt the design by supplying preconceived solutions, so it is inappropriate for the designer to preempt the construction process by specifying methods. Provided that the product being provided conforms to specification, how this was achieved is irrelevant. RAIA AN16.01.601 (1984) is quite forthright in its advice on this: "Instruct the builder as to the result to be achieved, not how to achieve it". Standen (1995) is ambivalent, seeming to advocate the use of methods on

30. Christopher Alexander (1964) calls this "fit and misfit".

the one hand but later advising that the contractor should be free to adopt any methods it pleases.

We have already seen (Chapter 5) that contracts can be explicit about this subject, so the specification of method may be in conflict with the contract. There are other reasons why designers should not specify construction methods, as follows:

- Lack of specifier expertise.
- Liability for failure blurred.
- Abrogation of responsibility, by designer and by contractor.
- Limitations on time/cost benefits available to purchaser.
- Creation of internal conflict between product and process descriptions.
- Need for supervision by contract administrator.

The contractor is expert on the processes and tools used in construction. The specifier is not. Just as the contractor may vaguely know why a product has been specified, so the specifier may vaguely know the processes that will be used to create the product. But it probably will not be a complete understanding e.g. in terms of how this process will best mesh with others. It will not necessarily be up to date e.g. it might require the use of equipment or materials that have been banned for health and safety reasons.[31] It may also not be the most appropriate for the project e.g. subcontractors with the appropriate skills may not be available at the time. The specification itself might be faulty if, for example, a critical step in the construction process was omitted. Specifiers have enough to do to become experts on specifying products, without having to worry about processes as well (though, of course, buildability must be considered).

Just as contractors can find themselves liable for failures against fitness for purpose when they design and construct, so specifiers might find themselves liable for failures against fitness for purpose where they specify both products and processes. The specifier should maintain a clean split between the pre-handover (design, under construct only) and post-handover (construction, under construct only) processes, otherwise liability for failure will become blurred. The specification should only describe the outcome of the design decisions. Construction decisions must await the contractor.

Contractors are in a position to advise that the specified method will not succeed, and may be found partly liable if they fail to do so, but to have this specification undone requires a variation and an embarrassing step-down by the contract administrator. This is all best avoided in the first place.

Hand-holding discourages contractors from thinking for themselves. When the specification of methods becomes traditional, as it is for curing of concrete or placing of asphalt, failure to specify method could spell trouble for the job. But methods are often specified when the products are not (as with curing), and this gives the contractor no choice but to follow the method (as a cook follows a recipe). The designer has ceased to

31. The ACSE Concrete Specification (1991), clause 12.3, states: "Any comment by the Superintendent as to the suitability of any erection system, or its particular application, shall not release the Contractor of any of his responsibility to safely carry out the work." A case of having your cake and eating it, or all care and no responsibility.

think in terms of the end product. There has been a systematic perversion of sound roles in procurement. Both parties have abrogated their responsibilities.

Specification of the properties of properly cured concrete is the right step, of course, with the ability to verify that they have been provided. The contractor is then free to select a slow but cheap method of curing, or a fast but expensive method, whatever is best in the circumstances (which may change as the job progresses). Being contractually bound to one method is unhelpful at best.

The contractors' skill, and competitive edge, lies in them being able to select freely a better method than their competitors. A desired result e.g. scratch-free chairs, can be achieved in many ways, including temporary protection, careful handling and storage, just-in-time delivery, even replacement of the odd scratched chair. Contractors should be free to select the method or methods that give a conforming result for the best price. It is not in the principal's best interests in terms of time and cost for methods to be specified. Principals, and those acting for them, want to allow contractors to be as competitive as possible, to give them room to compete freely at tender. This keeps tender prices down. The required result must be specified that much more tightly, but that is desirable for all sorts of other reasons as well e.g. quality assurance, or specifying organisation not involved in contract administration.

As we have seen (and will also see below), mixed specifying – of both the product and the construction method, in this case – can lead to conflict. To meet the product specification might require use of another method, or vice versa. Resolving the conflict requires an instruction from the contract administrator, and inevitably the contractor will have priced for the cheaper of the two specifications, so there will be additional cost.

Contract administrators are expected to check that the works conform to stated requirements. If these include methods, then the contract administrator must be on-site to ensure that these methods are being used (unlike products, processes are ephemeral). The contract administrator must supervise. And yet we are told repeatedly that this is not what the contract administrator does, instead "periodic observation of the work in progress" is provided. Specification of processes weakens these claims. Supervision of work specified by method will also mean treading on the toes of the contractor, subcontractors, foremen, and so on, creating bad feeling and disputes. It follows that specifiers must not specify methods.[32]

But consider the problem of buildability, which (including tolerances) accounts for around 15% of disputes involving specifications (Nielsen and Nielsen 1981). *No Dispute* (NPWC/NBCC 1990) recommendation 3.8 states:

Consideration should be given to the appointment of a 'constructor' in the design and documentation process.

32. And it follows from this that the specification of quality management systems, which is about procedures or methods, if you will, is also not permitted. The same goes for environmental management systems, and waste management (Gelder 1999).

Solving this problem only requires proper consideration of the construction process in design, it does not require the specification of it. Specification of method is tempting when new materials and components are being used, or when one does not trust the tenderers, or when specifiers are convinced they know best. However, these issues are most appropriately addressed through a sensible approach to pre-selection of tenderers, and through asking for method statements (non-contractual) to be submitted at tender. Non-contractual advice to tenderers on methods of construction is another possibility.

But, specifications commonly call up methods, if only by citing standards or manufacturer's recommendations. The British standards on workmanship (the BS 8000 series) are an extreme example, since they are only about methods in 'traditional' trades, quantifying 'good practice' (and so they are not cited by the UK's master specification, NBS). Sequence of operations, commonly specified, is one facet of construction methods, and may be specified under AS 4000 (1997). Statutory requirements for occupational health and safety may also require consideration and specification of methods (e.g. safe practice for roofing, and safe erection of steelwork).

This conflict between the need to sometimes prescribe methods and the contractor's right to select appropriate methods is difficult to resolve under traditional procurement regimes. Satisfactory resolution may require increased use of the design-and-construct procurement route, with the contractor preparing detailed documentation (including the specification). Whether the contractor would be wise to dictate methods to its (design and) construction subcontractors is another issue (and one not covered in this book).

Mixed methods

Building components are rarely specified using a single, pure specification method. Mixed methods are pretty well universal. For a cavity wall, for example (brands and standards have been changed to protect the innocent), we may have:

Proprietary:

- Face bricks: Blocko Pty Ltd, Snodgrass series, type F.

Reference:

- Mortar: Type M to AS 1234.
- Wall ties: Class C to AS 5678.
- Internal leaf: Class D clay bricks to AS 9101, perforated.

Performance:

- Cavity insulation: R-value 4 or higher.

Descriptive (assuming no drawing shows this):

Construction: 110 mm outer leaf, 50 mm cavity, 76 mm inner leaf, plastered on internal face.

Such specification text will usually be accompanied by complementary drawings showing assembly and various details. So, in principle, mixing is possible. But, the use of different methods gives various messages to the contractor in terms of latitude, responsibility, quality level and precision. Be sure that these messages are intended.

The danger of mixed methods is that the same element and/or attributes may be specified twice, using different methods e.g. performance and proprietary, for the cavity insulation. These descriptions may conflict, creating ambiguity and uncertainty as well as conflicting messages on latitude and so on. To quote an American construction attorney (in Stitt 1994):

... design professionals should think twice about combining final performance requirements with materials and [methods] specs. Not only do they lead to a false sense of security – since the performance requirement may not be enforceable – they have also generated successful contractor claims based on time and expense incurred in attempting to meet them after materials and methods requirements had been satisfied.

This particular combination is fairly common, as is the combination of referencing a standard and specifying a particular proprietary item. With no rule that the latter overrides the former, what is intended if the proprietary item does not comply with the standard? Specifying water-resistant plasterboard generically followed by a proprietary citation of a non-water-resistant product again leads to conflict. The only reason specifiers are not more aware of these conflicts is that often no-one has a close look at the specification until someone discovers that there is a problem with what was built. This is too late. To paraphrase Stitt (1994), architects pretend to write specifications, and contractors pretend to read them.

A good example of a mixed specification, in which the same thing is specified both descriptively and by performance, is that for Noah's ark, Genesis 6:

Make yourself an ark with ribs of cypress; cover it with reeds and coat it inside and out with pitch. This is to be its plan: the length of the ark shall be [135 m], its breadth [22.5 m], and its height [13.5 m]. You shall make a roof for the ark, giving it a fall of [450 mm] when complete; and put a door in the side of the ark, and build three decks, upper, middle, and lower. ... And you shall bring living creatures of every kind into the ark to keep them alive with you, two of each kind, a male and a female; two of every kind of bird, beast, and reptile, shall come to you to be kept alive (anon 1974).

There is potential for serious conflict between the specified size (of a large floating box, not a boat – there is no mention of a keel or sails and contractors should not read between the lines) and the performance requirement to contain a pair of every creature. The Bible goes on, however, to muddy the waters by describing a performance-based variation, Genesis 7:

Take with you seven pairs, male and female, of all beasts that are ritually clean, and one pair, male and female, of all beasts that are not clean; also seven pairs, male and female, of every bird ...

The potential for conflict was evidently realised with this variation, because we then observe that "... into the ark with Noah went one pair, male and female, of all beasts, clean and unclean, of birds and of everything that crawls on the ground, two by two, as God had commanded." Seven pairs of some creatures clearly did not fit! There must have been frantic negotiations, particularly given the penalty for non-completion, Genesis 7:

In seven days' time I will send rain over the earth for forty days and forty nights, and I will wipe off the face of the earth every living thing that I have made.

Non-written specifications

Quality of workmanship and materials can be very hard to define quantitatively in the specification but, under quality assurance and Abrahamson at least, it is vital that the specification is self-sufficient in this regard. The traditional 'to approval' approach is a cop out at the best of times, and is entirely inappropriate under quality assurance.[33] The solution requires lateral thinking. For example, the specification could include photographs of the range of acceptable workmanship, as in AS 3610 (1995) – the formwork code. It could reference specific parts of local buildings as setting standards of acceptable and unacceptable workmanship (a suggestion taken up with a vengeance in the PC-1 contract, at clause 9.1, discussed in Chapter 5).

Figure 6.4: Conflict in the documents

33. Hall (1994) notes that "although 'the architect's satisfaction' may be a treasured thing, the question is whether it is a commodity which a contractor can reasonably cater for in his pricing and execution".

Prototypes or samples could be included as part of the contract documentation as Mitchell/Giurgola & Thorp did for elements of the New Parliament House, Canberra, with tenderers being asked to match the workmanship with their own prototypes. Samples of this sort would be particularly useful for natural products such as stone or veneers, and for products such as terracotta tiles and bricks. (This was actually anticipated in AS 1225 Appendix B, now superseded.) If a picture is worth a thousand words, a sample must be worth ten thousand.

This approach is not used anywhere near widely enough. Having the contractor supply samples and prototypes is common enough, but this is perhaps addressing the same issue the wrong (reactive) way around. Design, including documentation, should be proactive.

Unknowns

In documentation for new buildings there is usually no excuse for treating work as unknown, the common exception being ground conditions, which influence excavation and design of footings. The uncertainties associated with ground conditions are usually dealt with using management clauses defining "rock" and "bad ground" (as in NATSPEC *Earthwork*), and procedures for determining quantities and for adjusting the contract sum e.g. using a schedule of rates supplied at tender. Commonly, a certain quantity of "rock" is assumed in the documentation. Extras, if any, are usually covered by the contingency sum rather than a provisional sum. The contingency sum should not be stated anywhere in the contract documentation (Standen [1995] suggests the opposite). To do so is to invite claims. The issue of possibly contaminated land also needs to be addressed in certain situations.

Conversely, in refurbishment documentation (including demolition) uncertainties abound. How much asbestos will be discovered? How many windowsills will need to be replaced and how many can be patched? What is the condition of concealed structural timbers? These are commonly dealt with using a schedule of rates, for likely items of work, and provisional sums, which may end up either not being spent at all or being overspent considerably.

Thorough pre-documentation investigation (e.g. condition surveys) is clearly essential to minimise the principal's exposure to such uncertainty. The provisional sums should only be scheduled in the *Preliminaries* or in a special schedule. This is to ensure that the subcontractors do not see them at tender.[34] The nature of the likely work (e.g. materials, workmanship) must be specified in the appropriate section of the specification (and/or drawn, as appropriate), and default quantities (if any) provided in schedules of work.

In practice, due to pressure of time, specifiers all too often cut corners by postponing design decisions or by failing to carry out adequate research e.g. on-site, for alterations and additions. The result is that more of the work is unknown than would otherwise need to be. This problem would largely be solved if the specification was begun on time. There also seems to be confusion over the distinction between prime cost, provisional and

34. This assumes that the Preliminaries have been stripped of material of interest to subcontractors, as NATSPEC has done in creating the General requirements worksection. This way only the contractor will see, or needs to see, the Preliminaries.

contingency sums. The conditions of contract do not help, by lumping them all together.

The use of provisional sums to delay design and documentation until the construction phase is not advisable. Administration of provisional sums is time-consuming and can become complicated, magnifying delays and leading to higher-than-necessary costs. Do not be a lazy specifier; it is counterproductive.[35]

Prime cost sums are used where a firm price has been obtained i.e. where a nominated (or identified) subcontract or nominated (or identified) supply contract has been tendered and agreed before the prime contract. Executing the work at this cost does not constitute a variation. The contractor will add overheads and profit to this amount, hence the term 'prime' cost (i.e. excludes overheads and profit). If the work is not done, the prime cost sum plus overheads and profit is deleted by variation.

Prime cost sums are used for identified items of work and, in terms of certainty, are better than provisional sums. Provisional sums, however, are used for unknown work, including nominated subcontract work to be let after the prime contract. If the work is to be done, a variation is needed (which first deletes the original provisional sum from the contract sum). Agreed labour rates and unit prices, including overheads and profit, will be used to cost the work, and this amount will be added back to the contract sum. If the work is not done, the sum will simply be deleted by variation. For provisional allowances, see Standen (1995).

Specifying mechanical and electrical services

Current specification practice in the mechanical and electrical services sector leaves much to be desired. There are three common methods of documenting services:

1. Specifying the overall system by performance. If the rest of the building exists or has already been designed, contractors will be given documents describing these 'givens'. The contractor designs the whole system, selects equipment, installs and commissions, and tests to demonstrate compliance with the overall performance requirements.
2. Specifying many pieces of equipment by performance, but reducing the design involvement of the contractor by the supply of a detailed design for the rest of the system. In particular, reticulated elements such as ducts, pipes and cables are specified prescriptively and are sized by the designer. To be confident that sizes are right, designers often assume that particular pieces of equipment will be used. The performance characteristics of these pieces of equipment, taken from the manufacturer's catalogues, are the (unidentified) basis for the equipment specifications. As stated above (under Performance specifications), specifying by performance presumes that the item so specified has to be designed. This is done to take advantage of the contractor's design expertise, which is felt to be superior to the specifier's. It is an abuse of this approach if the contractor is simply expected to find a complying product – often already selected by the specifier – in a catalogue.

35. In a 1996 NSW Supreme Court case (Australian Competition and Consumer Commission v. Collings Construction and Venture Industries), the judgement found that the contract price "was constructed on a contractual foundation of provisional sum items and provisions for variations, which, in the circumstances ensured that the contract price could be greatly increased during the performance of the works ..." (ACCC 1997). There is not much certainty in that!

3. Combining the first two approaches produces what is perhaps the most common method. The overall performance of the system is specified, but outline drawings and a comprehensive specification – by performance for many items of plant and equipment, and prescriptively for the rest, including reticulated elements – are also provided. The contractor is required to recalculate pipe and duct sizes to suit selected plant and equipment particularly fans and pumps, and then to install, commission, and test the system to demonstrate compliance with the overall performance requirements. The same quirk, regarding use of catalogues on equipment selection, applies as for the second method.

System performance satisfies the "five desirable features of specification methods" listed under Methods – Overview (in this chapter), subject to a couple of constraints. First, it can only be used under a design-and-construct contract, which may be for the entire building, for the services contract only, or for a nominated subcontract. As stated under Performance specifications, tenderers must be able to demonstrate that their proposal will meet performance requirements. This can be difficult, and submission of sketch proposals with computer simulations may assist.

The other constraint is with the rest of the building. Not much can be done about it if the building already exists, but if it is already documented as construct only, then an opportunity to properly integrate and coordinate the service system with the rest of the design has been lost. The system performance approach works best for services (which wend their way through the entire building) when the rest of the building is also design and construct, making it possible to integrate services. Otherwise plant rooms may end up being too small, ceiling spaces and shafts too tight, lift overruns inappropriate, power supply inadequate, and so on. Non-services designers can only anticipate these issues up to a point, particularly given cost pressures to minimise the number of non-lettable spaces such as those for services (and storage and stairs).[36]

Equipment performance, which mixes contractor-design and specifier-design, runs into problems straightaway. The only forms of contract in common use in Australia are either pure design and construct or pure construct only. Hybrids, of the sort alluded to in Table 6.8 are not available, but they are the required form of contract if this specification method is to be used legitimately. In fact, the contractor is not expected by the specifier to design the plant and equipment, but simply to select it or, more truthfully, to read the specifier's mind. However, the contractor, in selecting complying equipment, may find that the reticulated elements designed by the specifier need to be resized. Because the equipment and the reticulated elements are related so closely it is difficult to see who is responsible for this redesign (and therefore for the system design), so we have an internally inconsistent method of specifying, both in a technical sense and contractually.

Specifying equipment by performance is an attempt to get a good price, but it can

36. Integration is not such an issue for curtain walls and cladding, the design of which can readily accommodate given structural and other constraints, if the structural and other designers have borne the curtain walling and cladding in mind in their design work.

backfire quite easily –for example if there is only one complying product or supplier (not uncommon if acoustic criteria are tight, for example). Last, it becomes even more difficult to integrate services. Not only may associated reticulated elements need to be redesigned, but this approach is commonly used under a construct-only contract, which means that plant room sizes etc. cannot be changed. Selected equipment may only just fit, jeopardising easy maintenance, access, connection, testing and replacement. Other problems with integration, of the sort mentioned above (e.g. inappropriate lift overruns) can also arise.

The method that combines the two previous approaches, what might be called 'redesign and construct', is a mess. It is doubtful that even a hybrid form of contract could be devised to cope with it. After all, stipulating overall performance (not just 'for information only', but contractually) implies that the contractor is to design the whole system. Specifying a prescriptive/performance mix and providing outline drawings conflicts with this; clearly the specifier has done design work already. And yet, and this is the most bizarre part, the specifier implicitly concedes that the given design will not succeed. The contractor has to do detail design and perhaps redesign the whole system, which brings us back to where we started!

The contractor should, but cannot, rightly ignore the given design altogether – it only muddies the waters. Just who is responsible if the system fails to perform is unclear to say the least. The same thing has been specified in two almost certainly conflicting ways. The method is both contractually and technically inconsistent. The courts are likely to hold that the prescriptive requirements override the performance requirements, even if it is a design-and-construct contract. The attempt to pass design responsibility on to the contractor including any system redesign resulting from the contractor's equipment selection is flawed under construct-only contracts such as JCC or AS 4000. At least responsibility for redesign, unclear in the second method, has been addressed, even if unsatisfactorily!

This approach is disastrous for getting a good, or even a meaningful, price. For a start some tenderers will allow for redesign, but others will take a chance and not do so. Some will assume that if they select the plant as specified, the overall performance will be achieved automatically; others will not. Some contractors will appreciate that they may be responsible for overall system performance, while others will argue that this was given only as background information as an aid to detail design. On coordination and/or integration, this approach inherits all the problems outlined above, with the added question of where system design and redesign stops. For example, specification of several slightly oversize chillers, selected to an equipment performance rather than a system performance specification, may mean that the switchboard (and switchroom) has to be redesigned as electrical loads increase beyond expectation. This situation is not unheard of, but who is responsible for it?

So, the first method (system performance) is acceptable in some circumstances, but the second and third (equipment performance and redesign and construct) are fraught with difficulty. Do services specifiers have any other options for plant and equipment? Reticulation is well covered, with standards for ductwork and insulation available. Full descriptive specifications are out of the question. Chillers and the like are far too complex, and besides they are usually bought off-the-shelf or made up using a proprietary system or kit-of-parts.

For many items of mechanical plant and equipment e.g. cooling towers, water heating boilers, fans and pumps, there are no overall locally used standards. Hydraulics and electrical services are better served; but standards for central switchboards, more correctly low-voltage switchgear and controlgear assemblies, still need a great deal of supplementary material, for example. Even where there are standards, however, they still need to be supplemented with project-specific performance data to be useful. The only other option is specification by proprietary item.

As we have seen, this option is often used surreptitiously in any case, in the guise of a performance specification. Eminently suitable for much services equipment, this approach is legitimate only under a construct-only contract (still the most common type of contract). But it offers clear responsibility for design, assures internal consistency both technically and contractually, and permits the plant and its associated reticulation to be fully coordinated and the system to be fully coordinated and integrated with the rest of the project – and this is all done at the design stage.

In short, this approach only fails to meet one of the "five desirable features of specification methods" i.e. it does not obtain the best price. This is because a particular item will not be available through a number of suppliers, allowing contractors to shop around, but will only be available through one supplier – the manufacturer. Solutions to this have been outlined above (under Proprietary specifications), and include

- using nominated suppliers and subcontractors, tendered during the documentation program;
- listing a series of proprietary alternatives, from different manufacturers; and
- using existing supply agreements between the principal and manufacturer.

The first approach is common for lifts, where integration with structure and services is so important (affecting shaft size, lift pit size and structure, lift overrun, lift motor room configuration, pulley room, power, communications and so on). Mechanical and electrical specifiers are also accustomed to this approach, even though it seems to be declining in this sector. Indeed, many specifiers habitually specify as if this is the procurement method being used (hence "work by others" for example), even when it is not (see Chapter 5).

The second approach seems useful on the face of it, but is relatively new. It is not uncommon to list "Trane, Carrier or equal", which is as good as meaningless. The third approach depends on convincing the principal of the worth of such an arrangement. This may be most straightforward to do with a principal that has a great deal of construction work and associated maintenance, such as a public works authority or a university. In any event, it is not unknown.

Summary

Systematically resolve quality levels for the project and components (ideally, in consultation with the principal and contractor, hence design and construct), and communicate them unambiguously, both to the principal and to the contractor (and to other users of the specification). Communicating the requirements for quality is the main technical function of the specification.

Review specification methodology with an open mind. Many specifiers approach the task assuming procurement routes that were typical in the 1950s. Times have changed enormously, and so a fresh look at the task is needed. Customary approaches may no longer be appropriate.

Consider the extent of contractor choice in satisfying requirements implicit in the method chosen.

Ensure that the specification method is consistent with the contract. In particular, ensure responsibility for design is clear. Traditional contracts make little or no allowance for contractor design. Use performance specifications only if the contractor is to design.

Ensure that the specification method allows scope for obtaining a competitive price, either through pre-tender negotiations e.g. nominated subcontracting, or through shopping around by tenderers and contractors.

Ensure that the specification method allows for integrating and coordinating the component or system with the rest of the project. This is a major issue for services. The opportunity for efficient integration and coordination is lost if the services are designed by the contractor, but the rest of the building is not.

Specify the verification methodology e.g. tests, samples. This is part of the work to be done, and must be priced.

Specify the management of the specification method e.g. substitutions. In particular, major proposals for substitution should be made known at tender (which must be stated in conditions of tendering, not in the specification).

Use drawings in lieu of descriptive specifications, as far as practicable. They are generally clearer. Drawings (not the specification) should also be used for assembly instructions, since clarifying assembly is one of their main functions.

Be careful when using proprietary specifications – their simplicity is deceptive. Especially avoid "or equal". There are problems with closure, substitution, anti-competition and sometimes inappropriate use e.g. for aesthetically important items only.

Use reference specifications (performance or prescriptive) as far as practicable, but with care. Manufacturers work to them, and this method keeps the specification reasonably open, but compliance can be hard to demonstrate (third-party certification schemes are recommended) and familiarity hard to attain. Precise citation can be problematic, and there may be problems with currency, quality and relevance.

Do not mix methods for the same component or system. Ensure the specification method is internally consistent e.g. do not simultaneously provide system-wide performance requirements and equipment performance requirements as this may unduly limit the designers' ability to satisfy the overall design requirements.

Consider using photographs, models, samples, etc. Specifying quality can often be more easily done this way than by using words, especially for natural products such as stone. This may well be a way of coping with the quantifying quality dilemma posed by quality assurance and the like.

Minimise the use of provisional sums. In other words, minimise the unknowns and avoid procrastination. It is much more efficient to resolve the design proactively. This, after all, is what the documentation process is about.

7 Editorial issues

Production of specifications is a serious matter. As indicated in the Preface, they are of interest to many users. To ignore any single application – especially legal implications – is to invite disaster, particularly as the specification is often turned to only when things have gone awry, and perhaps to litigation. In law, the specification is at least as important as the drawings, sometimes (e.g. JCC) more so.

It is not good enough, however, to look at the specification only when something goes wrong. It would be unacceptable if a floor plan or a carefully resolved detail was not examined and followed exactly. Similarly, it should be unacceptable for the quality defined in the specification, on the principal's behalf and in the principal's interest, to not be understood and achieved.

For specifications to be used, they must be usable. On the technical front, specifiers must be sure they have specified items capable of satisfying project requirements, and have used the latest information. This chapter, however, concentrates largely on editorial skills: clarity, brevity, and logic,[1] and on central control.

Under Clarity, the chapter looks at word usage, ambiguity (hard for authors to spot), and repetition. Brevity is about the specification as a whole, and about the terse style used in modern specifications. Under Logic, the chapter examines structure, format, the right location for certain material, cross-referencing, and redundancy.

On central control (discussed under Editing), the final specification (final in terms of technical content) needs to be edited, tidied up, properly organised, well presented, and then centrally printed and distributed to interested parties and to a succession of users. As a contractual document it must also be controlled. For example, alterations to the original contract version cannot be permitted once the contract is let. And all tenderers must receive identical copies.

1. Main source for this information, and also for Equity in Chapter 5 of this book, is Rosen and Heineman (1990), Chapter 8 (which is replicated to some extent in RAIA Advisory Note AN14.03.100 [1994b]).

Clarity

Several court cases have touched on the comprehensibility of documents. In one, Hepples v. Commissioner of Taxation (1991-1992), a NSW Supreme Court judge held that a guarantee was unjust under the *Contracts Review Act 1980* because of its physical form, language, and the lack of advice provided as to its effect (ALLC 1996). All these things need to be considered in the preparation of accessible project specifications. There are a number of prescriptive principles that could be applied, judiciously, to achieve this performance objective. These are commonly lumped together under the banner of plain English.

The definition and use of plain English remains a controversial subject. The consensus is that it is not about simplification of expression (precision is still needed), and that it involves much more than the use of words. Visual, organisational, and sequential as well as linguistic aspects need attention. The definition supported by the Australian Language and Literacy Council (ALLC) makes this clear:

The use of language and design features so that a document is appropriate to its purpose, the subject matter, the relationship between the reader and the writer, the document type and the way the document is used (1996).

Its use cannot solve all communication problems, of course. And overuse can be counterproductive at the expense of precision – plain English can be taken too far.
The ALLC recommends

- writing clear, intelligible (not simplistic) English;
- making information more accessible (written to help the reader understand e.g. through keywords); and
- acknowledging the relationship between design and language choices, considering the purpose, nature of content, the reader-writer relationship (e.g. is the writer qualified to explain things to the reader?), and document type and use.

Typical plain English considerations include the following:

- Structure (logical organisation and accessible structure).
- The look of the document e.g. use of white space, typefaces, highlighting, explanatory notes, length of line, sentences and paragraphs.
- Language such as short sentences (about 30 words maximum), simple and direct modern English, no surplus words, no archaic legal expressions, use of active voice (verbs instead of nouns) and present tense (e.g. 'must' instead of 'shall'), positive statements (people find it easier to be told what to do rather than what not to do),[2] word order and general sentence structure, gender-neutrality, and consistent and precise language.

Most of these topics are explored in this chapter.

Avoid legal phraseology or highly stilted formal terms and sentences. Even in legal drafting, plain English or plain language tends to be used.[3] The specification is essentially

2. *Though negative statements are preferred sometimes e.g. No smoking!*

3. *For more on legal use of plain English, see ALLC (1996).*

a set of instructions from designers to constructors. If it is to be used on-site it must be understandable to users unfamiliar with legal styles, such as tradespeople and site supervisors, for many of whom English is not a first language.[4]

Some specifiers feel that what they are writing is not really a specification (i.e. a legal document) unless the style is stilted. The principles for writing specifications are different to those for books and letters but, at least since Peaslee (1939), they do not involve obscure, even archaic, language. A short list of words commonly found in legal documents, but not recommended for use, follows:

Above, abovementioned, aforesaid, and/or, below, deem, forthwith, hereby, herein, hereinafter, hereinbefore, hereto, heretofore, preceding, said, same, succeeding, such, whatsoever, wheresoever, whosoever.

You get the idea! A lot of these are to do with cross-referencing. Use of clause titles is preferred as will be discussed later. Other common words and phrases, with suggested alternatives, follow (Eagleson 1990):

adjacent to	=	*next to*
alternatively	=	*or*
approximate	=	*about, roughly*
as a means of	=	*to*
ascertain	=	*find out*
attempt	=	*try*
by means of	=	*by, with*
in case of	=	*if*
commence	=	*start, begin*
concerning	=	*about*
as a consequence of	=	*because*
construct	=	*make*
deem	=	*consider*
demonstrate	=	*show*
designate	=	*appoint*
discontinue	=	*stop*
employ	=	*use*
encased	=	*within*
ensure	=	*make sure*
establish	=	*show*
examine	=	*check*
facilitate	=	*ease*
following (in time)	=	*after*
for the duration of	=	*during*

4. *For example, the 1996 Census showed that 2.6 million people (about 15% of the population) spoke a language other than English at home. The biggest language groups, other than English, are Italian, Greek, Chinese, Arabic, German, and Vietnamese (in descending order). These non-English speakers are often localised. For example, in NSW over 50% of the population in Fairfield, Canterbury, Marrickville, Ashfield, Botany, Burwood and Auburn spoke a language other than English at home in 1991 (Jupp 1995).*

furnish	=	*provide*
identical	=	*same*
immediately	=	*at once*
in a timely manner	=	*promptly*
indicate	=	*show*
in order that	=	*for, so*
inspect	=	*check*
limited number	=	*few*
magnitude	=	*size*
maintain	=	*keep*
minimise	=	*lessen*
necessitate	=	*need*
notify	=	*tell*
not later than	=	*by*
obtain	=	*get*
otherwise	=	*or*
position	=	*place*
potable	=	*drinkable*
preclude	=	*prevent*
prescribed	=	*fixed*
prior to	=	*before*
regarding	=	*about*
relocate	=	*move*
remainder	=	*rest*
request	=	*ask*
review	=	*check*
said, such, same	=	*the, this, that*
select	=	*choose*
shall (obligation)	=	*must*
similar to	=	*like*
subject to	=	*as long as, if*
subsequent to	=	*after*
supplementary	=	*extra, more*
terminate	=	*stop*
therefore	=	*so*
until such time as	=	*until*
utilise	=	*use*
whenever	=	*when*

The courts often interpret terms in line with common usage (i.e. normal literal meaning), not the usage peculiar to the construction industry:

The purpose of interpreting contract documents is to ascertain the real intention of the parties as expressed in the words they have used. It is not a matter of speculating upon what the parties intended to write and words which the parties used will be taken in their plain ordinary sense. In searching for the intention of the parties 'no narrow or pedantic approach is warranted, particularly in the case of commercial arrangements' (Dorter and Sharkey 1981).

This is one reason why architects are advised to no longer claim to 'supervise' or 'inspect' or 'provide quality control'. These terms may be misunderstood. Rather they 'make periodic observations of the work in progress'(with quality assurance they may not even do that!). Similarly, architects should offer only 'opinions of probable cost', not 'estimates'. With this in mind, 'contract administrator' may be a better term to use in specifications than 'superintendent'.[5] Heineman (1995b) discusses common terms dealing with building modifications and with contracts, which are often treated as synonyms or misunderstood by specifiers (e.g. rehabilitate versus renovate).

So, use ordinary words in the ordinary way e.g. do not use 'dig' (i.e. by hand) if you mean 'excavate' (by whatever method the excavator considers appropriate, maybe by hand, maybe not). Do not use jargon. Use technical terms only if there are no equivalent everyday terms.[6]

Use a simple imperative (that is, mandatory) mood wherever possible: say, "Do this, do that" rather than "The Contractor shall do this". The specification is addressed to the contractor, and no other, after all.

Avoid the use of redundant reference paragraphs at the start of each section. In particular, 'The general conditions and preliminaries form part of this section of the specification' need not be used. It goes without saying. The contractor is responsible for all work and must read all documents together (as should the subcontractors).

Avoid ambiguity. All users of the specification must interpret it in the same way. This applies particularly to tenderers and parties to the contract. The best check for grammatical ambiguity is a 'devil's advocate' review by an independent colleague to highlight all the double or hidden meanings that must be eliminated. Ambiguity accounts for around one in eight specification disputes (Nielsen and Nielsen 1981).

The *contra proferentem* rule applied by the courts generally means that ambiguity is construed against the principal (the author of the specification). This is one reason why consensus forms of contract, and documents like Australian standards and *NATSPEC Reference*, are to be preferred to project-specific versions, having industry-wide rather than unilateral authorship.[7]

Use the same terms consistently throughout the documents. Do not use a galaxy of words and phrases for operating instructions. Use a limited range e.g. 'use', 'provide',

5. Nevertheless, 'supervisor' or 'superintendent' (rather than 'contract administrator'), and 'inspection' (rather than 'observation'), are commonly used, even in NATSPEC.

6. Some technical terms are everyday terms, of course e.g. osteoporosis and vasectomy.

7. See sidenote 4 in Chapter 4. This approach, however, has been questioned on the grounds that the principal has implicitly decided to run with the consensus text including inherent ambiguities, which could have been corrected. That is, the principal has in effect written the document. The actual parties to the contract did not negotiate contractual terms – it is a take-it-or-leave-it document.

'submit', 'supply' with each having a particular (defined) meaning and application. 'Supply and install' or 'furnish' would not then be used. Do not use synonyms (to avoid boring the reader). If you change the term you change the meaning. The drawings, too, should use the specification keywords. This requires careful coordination, as suggested in Chapter 5.

Check for typographic errors, particularly numerical ones. These account for around 1.5% of specification disputes (Nielsen and Nielsen 1981), though they are more widespread than this figure suggests. While spelling errors are not usually fatal (the reader can often deduce what was intended), numerical errors are. There is no way the reader can work out what was meant, which is one reason for cross-referencing by clause title, rather than number, as used in NATSPEC specification material (further discussed later). Tables often move from draft to draft, and symbols may transmute from (the engineer's) wordprocessing system to (the architect's) wordprocessing system.

Avoid imprecise words or clauses. For example the meaning of 'smooth' varies according to what it is being applied to (asphalt, concrete, stone, steel). It is more precise to state what tool is to be used to achieve the finish e.g. wood float, steel trowel, carborundum stone, sandblasting, polishing. Similarly, do not use colloquialisms. Precision is essential.

Use 'should' only for guidance clauses where there is no requirement for compliance implied; this is not common and is not recommended in specifications. Standards that use terms like 'should' or 'might' or 'can' need to be 'hardened' if they are to be referred to in specifications (or the BCA), as advised in Chapter 5.

Use acronyms and abbreviations only if they are well-known and generally accepted (such as NATA, BCA, PVC, mm, kPa), or define them in the text. The same applies to technical terms.

If alternatives are acceptable, they should be indicated clearly and without confusion e.g. "Select from the following:...". Do not use 'any' or 'either' when a choice is not intended. Use

- "Remove condemned materials ... on both sides of main entrance".

Do not use

- "Remove any condemned materials ... on either side of main entrance".

Restrict the use of "A and/or B". Instead use "A or B or both".

Use of different words to amplify or clarify a statement may instead reduce clarity. If a statement is not satisfactory it should be discarded and rewritten, not amplified elsewhere. Avoid repetition within and between the specification, drawings and other contract documents. Do not describe in the specification what is already described in equivalent terms on the drawings, and vice versa (e.g. brick leaf thicknesses).

These two documents are meant to be complementary. Conflict (a form of ambiguity) between plans and specifications accounts for around one in eight disputes involving specifications (Nielsen and Nielsen 1981).

Requirements may be repeated by mistake, or for emphasis, or in an attempt to be user-friendly. Whatever the reason, it is misguided. At best, it is a waste of resources that might be better used closing gaps rather than creating overlaps. At worst, it is dangerous. Whiteley (1998) agrees, stating that repetition "threatens consistency and concision", fundamentally disagreeing with contrary advice in Willis and Willis (1997) which states: "It may mean some duplication but this does no harm; in fact repetition helps to impress ...".

Reasons for avoiding repetition include the following:

- The potential for double pricing by tenderers, with little chance of a rebate from the contractor. This might arise with ducted skirting being specified both under metalwork and electrical worksections, or door grilles both in metalwork and mechanical worksections, for example. This is simply a result of carelessness.
- The potential for conflicting and inconsistent work practices and use of materials. Trenching specified in *Service trenching*, for example, may be quite different to that specified in *Hydraulics*. Ditto for insulation specified both under *Roofing* and *Cladding*, or painting both in *Pipework* and *Painting*. Which one applies in a single-prime contract project? Again, this is a result of carelessness.
- The potential for tenderers (especially subcontract tenderers) not to examine, or even sight, significant parts of the contract documentation. This underuse of documentation, which leads to unrealistic tenders and all the difficulties that flow from this, might arise in several ways, all common in practice, and usually intended to enhance user-friendliness:
  - Extensive notes on the drawings. Why read the specification when all the 'important' stuff is on the drawings? Why take the specification seriously when it is clearly redundant?
  - Extensive 'scope of works' or 'outline of works' statements. Why read the rest of the specification when you are a tenderer in a hurry and an apparently comprehensive extract is available (common in services worksections)? These statements are also often of dubious contractual status.
  - Repetition of material in statutes or cited standards. Why read the statute or standard in detail when all the 'important' stuff is in the specification? Why take all those other documents seriously? Further, simply citing the odd statute is dangerous. It carries the implication that other statutes are less important or do not apply.
  - Provision of bills of quantities in tendering. Why read the specification or examine detail drawings when all the information you need for pricing is in the bill? Bills, valuable in many circumstances, are also discussed below.
  - Provision of hold and witness point or inspection/test summaries. These may also discourage closer reading of the specification (by the contract administrator as well as the contractor).

- The potential for ambiguity due to errors in repetition (usually of the user-friendliness variety). These errors will generally be construed against the principal, with the court opting for the lowest quality level described, for example. Errors can arise in many ways, as follows:
  - Changes during documentation, without correction of all citations (the one that got away). This is common with notes on drawings, with drawings that have not 'jumped' scales,[8] numerical cross-references, and with finishes, often both scheduled in the specification and included on annotated plans, sections, reflected ceiling plans, etc.
  - Repetition that is inexact and capable of a different interpretation (especially by contractors looking for excuses).
  - Paraphrasing or rationalisation of standards may be in conflict with, or at least capable of a different interpretation from, the original standard, which may also be cited.
  - Even precise extracts from standards in the specification may have been superseded by amendments or new editions, which may also have to be used according to a *General requirements* clause, for example.
- The implication that non-repeated material is less important or less critical. If it is in the specification or on the drawings, it must be important. To imply that certain parts are more required (for emphasis) than others is dangerous.

There is an argument that redundancies are useful to help readers from non-English speaking backgrounds unambiguously grasp the specification's meaning, but while this may be acceptable in census forms, insurance policies and instructions for use, it is not appropriate in a specification.

Although documentors clearly intend that tenderers and the contractor use all documents, in trying to be helpful they may undermine this goal. The classic example is the use of bills of quantities, which directly conflicts with the principle of non-repetition. From this point of view at least, contractual bills in particular should be avoided (contrasting with other points of view that bills should be provided e.g. to assist in contract cost management). By all means prepare bills for cost management by the principal, but the best way to ensure that tenderers, and contractors, take full responsibility for their own estimates, is for them to prepare their own bills. Not everyone would agree with this point of view, of course (Davie 1995a, 1995b; Gelder 1995a; Poole 1995).

Bills contain no new information. They merely repeat in a different (convenient-to-estimators) form what is already in the specification and drawings. They inherit errors in the parent documents (while usefully identifying many during their preparation), and inevitably contain new errors themselves, euphemistically known as undermeasures or overmeasures. In fairness, it should be pointed out that the main sources of errors in the bills are the documents provided to the quantity surveyor. The widespread practice of preparing bills from incomplete specifications and drawings only makes this situation worse.

8. This refers to the recommended practice of avoiding 'consecutive' scales in the drawings i.e. avoiding neighbours in the series 1:1000, 1:500, 1:200, 1:100, 1:50, 1:20, 1:10, 1:5, 1:2 and 1:1. Drawing plans at both 1:100 and 1:50, for example, provides little extra information, but guarantees repetition and creates unnecessary work. It is more profitable – more efficient – to 'jump' scales e.g. provide sections at 1:100 and 1:20, skipping 1:50.

Lack of cross-referencing to the specification and drawings undermines the usefulness of those documents for tenderers, who will often work happily with the bill only (see Chapter 5). Often the bill is all that the contractor gives subcontractors for tender purposes, in any case. Indeed, the widespread use of repetition – between drawings, between specification and drawings, and between bills and specifications – tempts the contractor to simplify document dissemination by ignoring any documents that appear to be redundant, particularly the specification.

Brevity

Know when to stop. There is no such thing as 100% documentation but there is such a thing as over-documentation. The more you specify, the greater the risk of error and conflict, and the more responsibility you take. General conditions of contract contain catch-all defaults covering a certain amount of silence in the documents. Specifiers are not generally expected to cover everything.

The converse is under-documentation, which is also to be avoided. Ensure that everything required is included somewhere in the contract documents. Being overly concise can lead to ambiguity, so be careful e.g. "Doors: To AS 2688" is not enough. Options must be selected.

A common reaction from inexperienced specifiers is, if in doubt, leave it in! This reflects a lack of confidence and creates bulky documents. Leave it out! NATSPEC, for example, expects half of the prompts and reference material provided will not apply to a given project. If you do not understand a clause, you will not be able to administer it.

Few architects or engineers would feel happy about issuing a set of drawings that contain details not applicable to the project. Unfortunately there are some who do not have the same compunction about irrelevant material in the specification. Blanket unqualified reference to NATSPEC material is, amazingly, not unknown. Cull the specification as ruthlessly as you would cull the drawings.

Avoid lengthy verbal descriptions. Draw it or schedule it instead.

In phrases, do not make 'the contractor' the subject of a sentence unless this is absolutely necessary. The entire specification is addressed to the contractor (not the subcontractor or tenderers), after all.[9] That is, use:

Provide light fixtures ... (imperative mood)

Do not use:

The contractor shall install light fixtures ... (indicative mood).

In specifying materials, avoid 'shall', 'will', 'must', 'is to' and 'is to be'. Where such a structure is unavoidable, 'must' is preferred to 'shall' (Jackson 1993) and is used in the BCA (but not in standards). Do not use infinitives. Use:

Provide burnt clay bricks ...; or

Bricks: Burnt clay ....

9. General conditions of contract, on the other hand, have to distinguish between the responsibilities of several parties, so they have to use 'the contractor' etc. to differentiate.

Do not use:

Bricks shall be burnt clay ... (when?)
Bricks will be burnt clay ... (when?)
Bricks must be burnt clay ... (avoidable phrase structure); or
Bricks are to be burnt clay ... (when?).

Use streamlining, in the NATSPEC style. It is not new – it was first suggested 60 years ago[10] – and is widely used in national master specification systems all over the world. The style relies on colons and keywords thus:

Galvanizing: To AS/NZS 4680.

Do not use expressions such as 'all' or 'the whole of' where these are superfluous. Use:

Use new bricks.

Do not use:

All bricks shall be new; or
The whole of the brickwork shall be in new bricks.

Use 'which', 'who' and 'that' sparingly, if at all. The same applies to 'the', 'a' or 'an' and, even, 'of' e.g. "Paint finishes schedule" not "Schedule of paint finishes" (the former has a less legalistic feel to it). Careless use of 'the' or 'a' may restrict interpretation and application of clauses, possibly such that the strict interpretation will not be what you had in mind.

Use plurals consistently. Use of the singular may be taken to imply that there really is only one light fitting, even though the drawings show hundreds, for example.

Use numerals instead of words for numbers, except where one number follows another. Use:

Provide 6 copies ...; or *Use five 10 mm bolts ....*

Do not use:

Provide six copies ...; or *Provide six (6) copies ....*

Do not use 'same' as a pronoun or 'said' as an adjective. Use:

Replace rejected materials.

Do not use:

If materials are rejected, the builder shall replace same; or
Said materials shall be replaced.

Logic

Arrange information so it can be easily found. As stated in Chapter 4, the specification is structured primarily for navigation purposes. Separate work into worksections and subsections using a system with a logical search pattern, such as trades, components and construction sequence.

10. In 1939, the American architect Horace Peaslee advocated trimming unnecessary verbiage from specifications, so that requirements would be expressed "in clear, concise form in headings and subheadings, without sentence structure, using phrases in preference to clauses, with only essential adjectives or adverbs, with no articles, definite or indefinite, unless positively required".

Break the specification into small worksections, and standardise the sequence and titles of clauses and subclauses in each section. Arrange sentences in a logical order e.g. follow the construction sequence. Small project specification worksections help the contractor in the process of assembly or disassembly for subcontractors, just as small master specification sections will assist specifiers.

Break long clauses into subclauses, paragraphs and subparagraphs with titles or bullet points, in a logical hierarchy, each dealing with one item. Standardise text of common clauses and subclauses.

Provide titles for all clauses, subclauses and tables, for cross-referencing. Emphasise these for easy identification (e.g. bold, capitals). Refer to tables by their titles, not "... to the table below".

For points, or information, to note, use either lists (preferred) or expanded sentences, consistently. Lists are set out thus:

... including the following:
- This.
- This.
- This.

Expanded sentences are set out thus:

... such as
- this;
- this; and
- this.

Do not include text best dealt with elsewhere. In particular

- do not include in individual technical worksections material which is, or should be, covered in preliminaries, annexures to general conditions or the general conditions themselves e.g. project description and list of drawings; lists of selected subcontractors; inclusion of defects liability period, separable portions, operational maintenance period; management of valuations and discrepancies (Bennett 1980);
- do not duplicate material better described in other technical worksections e.g. trenching, painting;
- do not address tenderers or the principal or subcontractors; and
- do not include matters that do not relate to the contract.

Use:

Submit a works program at the first site meeting.

Do not use:

Submit a works program with the tender.

If the latter is required, it should be stated in the conditions of tendering. Other examples are

- tender schedules in electrical or mechanical worksections;
- *Allow for the inclusion of ..., Allow the provisional sum of ...*;
- *work by others* in single-prime contract specifications; and
- loose furniture schedules, where the contractor is only providing fixed items.

Be logical in what is said and not said. For example, do not

- state requirements but then nullify them by a statement indicating that a lower standard may be acceptable;
- state requirements and then state what will happen if they are not met – this is for the contract administrator to resolve as circumstances dictate; and
- use *Take special care to ...* or *Comply strictly with ...* and so on. These phrases, along with capitals, underlining or emboldening text, imply that some specification instructions are less stringent than others.

Minimise cross-references between sections and between the specification, drawings and other contract documents (see also Chapter 5). When using cross-references (e.g. if a related item is specified in another worksection) refer to clause (or worksection) titles rather than numbers because numbers are more likely to be changed during preparation of the specification. Use:

Refer to the Service trenching worksection.

Do not use:

Vide clause 4.13 on page 24.

This question of change also applies to the details, traditionally identified by number rather than by name. So be especially careful when referencing particular details. Review before issuing, or use titles rather than numbers, at least for the details themselves. Another exception is the citation of standards (their titles are often overlong), about which extra care should be taken. Numbers (section, clause and page) are essential for navigation purposes, however (e.g. over the phone), and should be provided.

There is no need for internal cross-references to have the meaning "conform to", since the contractor has to conform to the contract description (given in the documents defined in the contract appendix) in its entirety. If "Refer to" directs the contractor to external documents (i.e. not included in the contract), then the correct interpretation is surely that such a reference is for information only. If the intention is that the contractor conform to these external documents, then use the forms "Comply with", "Conform to" or simply "To AS 1234" (see discussion above on phrases).

Do not use the very unhelpful phrase "unless otherwise specified" (it may not be otherwise specified, but you have to read the whole specification to find this out). The use of the so-called 'residuary legatee' system is a better way of ensuring that every part of the project is covered. This relies on listing all particular requirements in one spot, and then the general requirement that is to apply in all other cases.

For example, use:

Toilet rooms: Obscure float glass.
Entrance doors and side lights: Toughened glass.
Borrowed lights: Plate glass.
Other locations: Clear float glass.

Do not use:

Glass: Clear float glass, unless otherwise specified.

Do not inadvertently include irrelevant material either from master specifications (e.g. instructions to specifiers, or text that does not apply to the project) or from previous project specifications. The document becomes unduly bulky, causes confusion, and will embarrass the contract administrator, undermining his/her authority. This sort of inclusion also undermines the authority of the document.

Do not include redundant material intentionally e.g. "Comply with the *Occupational Health and Safety Act.*" The contractor is required, by the general conditions of contract (or as an implied term), to comply with applicable legislation. A partial, possibly inappropriate, and possibly inaccurate, list is of little use and may mislead.

Finally, many of the standard phrases mentioned, that appear all through specifications, should be avoided simply because they are redundant. These include "unless otherwise specified" (already discussed), "the contractor shall", and "or equal". If you think about it, these could be attached to virtually every sentence (e.g. "The contractor shall, unless otherwise specified, provide masonry to AS 3700, or equal"). Including them with some sentences, but not others, is not logical. Attaching them to all sentences is patently unnecessary. The logical solution is to not include them at all (e.g. "Masonry: To AS 3700"). They generally go without saying, or are covered by clauses in the contract, *Preliminaries* or *General requirements* worksections.

Editing

Central control of the specification by the lead specifier, is vital. *No Dispute* (NPWC/NBCC 1990) recommendation 3.2 states:

A single point of responsibility should be allocated to ensure that documentation is properly co-ordinated.

This advice applies as much to the specification as to any other component of the documentation because, like the full set of contract drawings, the specification has been prepared by designers working in various disciplines, usually in separate offices. Individual contributions will inevitably be somewhat inconsistent, even if all contributors follow the advice given previously.

Coordination includes editing of the preliminary project specification, including services engineering worksections, by the lead specifier e.g. the architect. The editing function is vital for the specification's contractual integrity, which the architect would generally be in the best position to assess. It is also important to ensure a consistent and correct approach to technical and editorial matters (e.g. is excessive performance specifying used under a non-design-and-construct regime?), including simple readability.

Important editorial functions include checking for gaps and overlaps e.g. trenching in *Service trenching* and in *Hydraulics*, and simple coordination (in the manner of drawings coordination) e.g. if the building's frame is steel, are appropriate fixings specified for mechanical and other services? The editor does not have to be omniscient technically to edit, so technical ignorance of services engineering, for example, is not a valid excuse for shirking the task of editing that material.

And yet, architects often believe it is impossible to edit services specifications. The material is supplied too late, the architect does not feel technically competent to review it, and the editorial and contractual difficulties may seem insurmountable. This is one reason why services engineering specifications are still frequently so unsound. Services engineers have never received any constructive criticism from the architect. Increasing use of national master specifications such as *NATSPEC SERVICES* should help, despite widespread reluctance among services engineers to use such masters (better the devil they know?).

The editing process can be simplified if production is controlled by the lead specifier throughout the whole specification process, with proper briefing of the co-consultants' specifiers and a review of the drafts at each stage. It is essential to ensure that these people are fully informed about the mode of procurement (so that there can be no confusion regarding 'others', for example), the form of contract, content of the *Preliminaries* and umbrella worksections, if any. Policies of the principal that may affect the work of co-specifiers also need to be passed on by the lead specifier. These may mention use of proprietary specifications, for example, or preapproved suppliers and the like.

Prevention of problems is always better than a cure, particularly in terms of risk management and simple efficiency (one reason why the specification is written in the first place). The use of master specifications and computerised specifications has not removed the need for great care in the preparation of project specification material. Indeed, at least to start with, national master specifications require more care than most other sources since the specifier is neither familiar with their contents, nor with their method of use. All specifiers are likely to write their first master-based project specification at some point, and time has to be allowed for this learning curve.

Numbering system

For navigation e.g. when discussing the specification over the phone, the specification and drawings need a consecutive, logical numbering system. This is another aspect of documentation coordination.

There are many options for numbering. NATSPEC products use the following system: worksections are not numbered (the specifier is expected to do this), but the subsections within each section are (1, 2, 3 etc.). The specifier can add worksection numbers, so the subsections become 1.1, 1.2, 1.3 etc. Worksections could also be titled alphabetically (A, B, C etc.).

Clauses within each subsection can also be numbered, as follows: 1.1, 1.2, 1.3 etc. If worksection numbers (or letters) have been added, clauses will have three numbers e.g. 1.1.1, 1.1.2 etc. If subclauses are also to be numbered, a fourth level is needed i.e. 1.1.1.1. This is getting a little messy for a simple specification, but is perhaps warranted for complex documents. The electronic version of NATSPEC provides for automatic numbering for subsections and clauses, and use of master files.

Most master specification systems have a classification system for ordering their worksections, at least. For a number of systems this coding extends down to clause level. Classification systems are in a state of flux, both nationally and internationally. Systems in use in Australia include the NATSPEC classification system (*Classification2*, published in draft form, but used by CIS), the American MasterFormat system, the original (Swedish) SfB system, and its British development CI/SfB.[11] Most product database systems use their own. Clearly the scene is very confused. It is even worse internationally where many more systems in use (see Chapters 2 and 3).

Remember, the numbering system should not be used for cross-references. As already suggested, use words instead because they are more foolproof and resistant to change.

11. In the UK, the CAWS system (see also Chapter 3) is widely used for both specifications, and for drawings and bills as part of the Coordinated Project Information (CPI) initiative launched in 1987. On the drawings we might find a note "Tapered arch bricks [F10/123]", where "F10" is the brickwork section of the specification under CAWS, and "123" is the number of the clause titled "Clay bricks for arches". The specification, in turn, may have "Special shapes: As drawing C(21)15" where "C" indicates a component drawing, "(21)" indicates the CI/SfB element code, and "15" is the unique drawing number (RIBA, ACE et al 1993). (Uniclass has adopted CAWS, plus other systems for other entities.)

Contents and index

Provision of a series of contents lists will help readers understand – and follow – the structure of the specification. Provide a list of worksections at the beginning of the specification, and a contents list (subsections, clauses) at the beginning of each worksection no matter how small. If expectations for the contents of a given section vary among users, it would also help to provide a list of cross-references to related work described in other worksections. This is not the same as 'work by others'. The related work in other worksections may or may not become a part of the work of the subcontractors using this worksection.

Upfront contents lists are the norm, but lists for each worksection are not. However, they are more useful and less contractually dubious than a 'scope' clause, so should be used instead.

Another navigation aid is an index of keywords e.g. inspections, fasteners, galvanizing. This is included for information at the end of the specification, even if unlikely to be comprehensive. Specifiers can use it to detect repetition in the specification. Such an index allows the contractor to check globally for all inspections and tests and other quality control matters, for example, which helps with project planning. Authorities can also track down compliance issues this way. Those indexes included with *NATSPEC Reference* material provide a useful starting point for content. But indexes are hard to do, and time-consuming. They are rare in project specifications, though use of computers permits diligent specifiers to prepare project-specific indexes.

Supply of specifications electronically could allow searching of the document, which would be very helpful.

Appearance

Never underestimate the importance of appearance, neither for drawings nor the specification, as another aspect of document coordination. A scrappy specification, mixing typefaces, page layouts, footers and headers and so on from worksection to worksection (depending on who prepared it), implies hurried preparation and resultant lack of care and is not going to inspire anyone with confidence.

Even a fully consistent specification may fail in terms of appearance, despite the increasing ability of computer systems (Windows, Macintosh, UNIX, etc.) to exchange formatted text files (in the various editions of Word and WordPerfect etc.). The tests are legibility and, in particular, comprehension. The objective of the specification is communication, and not pursuit of fashion. The specifier has gone to a lot of trouble to research the contents of the specification and it is not worth jeopardising its usefulness for the sake of some uninformed concept of typographic design. Avoid typography that intrudes between the author and the reader; there are sensible typographical guidelines which should be followed such as Wheildon (1990), the source for the following advice.[12]

Colour is usually not an issue for specifications, which are generally black text photocopies on white paper. This is just as well, as the use of colour for both body text and headlines has a negative effect on comprehension. Besides, given the often poor lighting conditions on building sites, maximum contrast between text and background seems desirable. Black text on a white background is best in this respect, so do not use coloured paper. And do not use white on black – this is a total disaster!

Columns of text can be too wide or too narrow to be read and fully comprehended. Reasonable upper and lower limits seem to be 60 and 20 characters respectively per line (Wheildon 1990). Eagleson (1990) suggests 70 and 50, and Cooper and Page (1989), 65 and 35. Most specification pages are set out in single column format, so you should work to a maximum of 60, and leave a clear margin – handy for annotation in any case.[13] Ample use of white space is recommended anyway, to make the page less daunting. Spacing between clauses, subsections and so on gives visual cues to structure.

12. Wheildon's advice is good but it is based on newspaper reading.

13. For every rule there is an exception. This book has about 80-85 characters to the line!

Few specifiers attempt to justify the text both on the right and left. It is usually left-justified, and right-ragged. However such justification is worth doing, and is not hard with modern wordprocessing systems. Tests indicate that fully justified text maximises comprehension. Ragged-right text, the normal for specifications, halves this level of comprehension according to Wheildon. Yet it is *easier* to read, Eagleson says, as one is less likely to get lost from line to line; Cooper and Page agree.

Serif (Roman) typefaces such as Times New Roman[14] or Corona for body text type are likely to lead to two-thirds of readers attaining good comprehension. Sans serif typefaces such as Helvetica, beloved of architects, have the opposite effect with two-thirds of readers attaining only poor comprehension (Wheildon 1990). On the other hand Cooper and Page (1989) say that adults find little difference in legibility between serif and sans serif for body text. Clearly the safe option is to avoid sans serif typefaces for the body of the text. NATSPEC has taken this to heart in its *NATSPEC Reference* book and its digital material, but many architects reformat the disks to the ubiquitous (since 1957) Helvetica. This is counterproductive. Helvetica may be attractive and it may work for signage, but it does not work for a page of text.[15]

Avoid upper case body text too. Surveys found it to be almost universally regarded as hard to read, even in relatively small amounts (Wheildon 1990). The visual cues provided in lower case by the lettering's sticks and tails are missing.

Use of bold for blocks of body text halves comprehension, so avoid this. Surprisingly, italic text is only slightly less comprehensible than its Roman counterpart (Wheildon 1990).

Research on leaflet design indicates that 11 point typefaces are most widely regarded as easy to read, with 77% preferring natural 11 point, compared with 72% for 12 point, and 69% for 10 point (Wheildon 1990). Eagleson, meanwhile, suggests 10 or 12 point. Increasing line spacing (the leading) by one or two points helps considerably. Of respondents, 25% considered 11 point text on a 13 point body to be the most comfortable to read, followed by 19% for 10 on 12, and 18% for 12 on 13 (Wheildon 1990). Cooper and Page suggest a leading of about a quarter of the type size e.g. 10 on 12.[16] Given the ageing working population, and the usually poor lighting conditions on site, perhaps 12 on 13 is best for specifications. Transmission of documents by facsimile and electronically will also influence choice of point size and typeface.

Legibility is high for most Roman and sans serif typefaces used for clause titles, but it is higher for lower case than for capitals e.g. for Times Roman, 92% versus 69%, but only slightly higher for Roman than for sans serif (Wheildon 1990). Eagleson suggests that where two typefaces are to be used e.g. one for text and one for titles, they should contrast each other e.g. serif versus sans serif, bold versus light, capitals versus lower case.

Natural kerning (spacing of letters in a word) leads to maximum legibility. Avoid

14. Times New Roman first appeared in 1932.

15. Helvetica has been described as having "all the aesthetic qualities of a suburban terrace house: respectable, inconspicuous – part of the environment, in fact" (Klein et al 1991).

16. Caution – this advice relates to newspaper serif fonts, commonly Times. For a given point size, the physical size on 'x' height may vary considerably.

extreme kerning, where the letters touch; 90% condensed type i.e. narrower than natural, for clause titles achieves maximum legibility. If you must save space, condense a little, and do not kern.

In clause titles, full stops tend to pull readers up suddenly, reducing comprehension substantially.

Production

A few points need to be considered here, such as

- will the document be a single volume (a function of size)?
- will it be bound or loose-leaf?
- will it be single- or double-sided?
- will it be printed in-house or externally (a function of specification size versus reproduction facilities)? and
- will it be controlled[17] or non-controlled e.g. will it be maintained, can it be copied, will it be delivered electronically? [18]

Different versions may be prepared for different purposes. Tender sets will often need to be returned to whoever has put out the tender, but tenderers will need to copy them first for subcontract and supplier pricing. The sets could be loose-leaf to assist with copying and maintenance, with issuing of new sheets during the tender process to correct errors and discarding of old sheets. They could also be single-sided, and supplied in (recyclable) binders. These need to be controlled with a code on every page of the tenderer's copy, so photocopies can be traced to source. This may assist with identification of dodgy tender practices (e.g. collusion), for example.[19]

The two contract copies will be bound and controlled (marked or coded, for example), and can be double-sided (they will not be copied).

Sets used on-site can be loose-leaf, for copying, and controlled (but they will not be maintained – their contents cannot be changed). Depending on copying facilities, the sets can be double-sided (saving trees), but note that this affects footers which normally have the page number on the outside edge, and headers which would have the worksection title on the outside edge.

Double-sided versions should always start each worksection on the right-hand leaf, necessitating the occasional blank page at the end of the preceding section. This must be numbered and marked as being blank so it is clear that this is intentional and that nothing is missing. It is probably not worth having two versions i.e. one single-side and one double-sided.

Sets supplied to the facility manager should be loose-leaf, along with the user's manuals and other documentation handed over at completion. The facility manager or principal may also require an electronic version.[20]

17. Document control is a quality assurance requirement (see AS/NZS ISO 9001 [2000] for example). On what this entails and how it works, see RIBA Indemnity Research (RIBAIR 1989) and AS/NZS 3905.2 (1997).

18. See RAIA Advisory Note AN11.02.100 (1994a) on electronic drawings and document control, and Ellickson (1995).

19. A number of offices give each issue (i.e. document is sent out of the office) a unique issue number, with margin note numbers on the left against each change. This is very useful for informing team members about changes. NATSPEC and the BCA adopt a similar tack for flagging updates, with a sidebar, date plus a footer and/or header identifying the issue for the particular sheet.

20. Public works authorities often require this in a particular format e.g. Word for Windows, which is consequently very popular among specifiers and rapidly becoming a default among NATSPEC subscribers at least.

Distribution

Refer to relevant local legislation to find out how many copies of the specification (and drawings) should be sent to the local authority i.e. council. In the ACT, Western Australia and Tasmania, two copies should be sent; in Victoria and South Australia, three; and in NSW, four. A specification as such was not required in the Northern Territory.

Who holds on to the contract copies? The contractor holds one copy, and the architect commonly holds the principal's copy for the duration of the contract, returning it at practical completion. The principal can of course hold its own copy if preferred.

How is signing organised? Normally the two copies are first sent to the contractor for signing and sealing, if required (execution of the contract), or this may be done in the architect's offices to ensure it is done correctly. The executed copies are then sent to the principal for signing. One fully executed copy of all the documents is then delivered to the contractor, ideally by hand. Sometimes both parties attend a grand signing ceremony, making distribution convenient for the architect.

Other processes that should be addressed in the tendering material sent to tenderers include

- how many copies are supplied to each tenderer?
- what about extra copies? and
- do the tenderers collect them, or should you, the specifier, send them – and how (by registered mail)?

After tendering, how many copies are supplied to the contractor at no charge, and how many can be charged for? This is dealt with in SBW-2 clause 2.1 (three sets are supplied free), JCC clause 2.05 (10 sets initially, and 10 further sets), PC-1 clause 6.1 (as scheduled in the Contract Particulars) and AS 4000 Annexure Part A Item 15 (five sets, or as scheduled), for example.

How much should you charge for extra copies? RAIA Advisory Note AN15.01.100 (1993c) suggests current commercial rates.

Should you supply copies direct to subcontractors? The short answer is, no. What about copies to the contract administrator, project manager, co-consultants to assist with running the job? Which will be the 'master mark-up' copy, to record modifications as the job progresses on-site? How many copies should be sent to the facility manager? Probably the same as the number specified for user's manuals etc. i.e. three in NATSPEC *General requirements*.

Is a record copy required and if so, who will prepare it? What about archiving, including by co-consultants? A single copy for all design consultants is probably all that is needed. Others can always be copied if necessary. Do not be tempted to archive the specification or any other documents electronically. Hardware and software are evolving so fast that already archivists are facing serious problems with accessing old computer records (Rothenberg

1995; Achenbach 1999).[21] The documents should be held for at least seven years after completion, to cover possible claims for breach of contract or negligence.[22]

Summary

Be clear. For example, use plain language, not 'legalese'. Do not give special meaning to ordinary words. Do not use 'supervise', 'superintend' or 'inspect'. Use the imperative mood. Avoid ambiguity by, for example, using words consistently and precisely. Do not use 'should'.

Do not repeat material, whether by mistake, for emphasis, or in an attempt to be user-friendly. Say it once, and in the right place. This avoids double pricing, inconsistent work practices, giving the impression that some documents are redundant or less critical, and possible conflict. Use bills of quantities with great care, if at all.

Be brief and concise e.g. use streamlining. If in doubt, leave it out. If possible draw it, do not describe it in writing. Rephrase to avoid using 'shall', or use 'must'. Do not make 'the contractor' the subject of sentences.

Arrange material logically, for navigation. Break the specification into small worksections. Break long worksections, long clauses etc. into smaller units. Minimise cross-references. Where required, use keywords (titles) not clause numbers. Do not use "unless otherwise specified". Do not include irrelevant or redundant material e.g. from previous project specifications, and do not include text best dealt with elsewhere e.g. in *Tendering*.

Edit and coordinate the specification centrally (by the lead specifier). For this to be possible, the various parts of the specification must be ready on time. The extent of this editing can be minimised by proactive briefings and analyses during the documentation phase.

Provide a rational numbering system for navigation but not for cross-referencing. Provide a general contents list, and worksection-by-worksection contents lists. Consider providing an index, and consider providing the specification electronically.

Present the specification (format, typeface, layout) to optimise communication and comprehension. In particular, use serif typefaces, perhaps 12 point on 13.

Print and distribute the appropriate number of copies, in appropriate formats. Consider document control, and consider preparing a project record specification.

Up to a point, most lead specifiers, commonly architects, are serious about certain aspects of finishing the specification, including numbering and appearance. The chief stumbling block, for these and other aspects of finishing, is the material supplied by other specifiers. While problems with software compatibility are receding, 'Just-in-Time' delivery of the material remains an issue. This applies to drawings as well. It is often too late to do anything with it.

21. Achenbach says, "From a preservation standpoint, an electronic document may as well have been scrawled with a wet finger on a paper napkin".

22. RAIA Advisory Note AN11.01.101 (1998c) recommends this. AN02.01.100 (1998a) also recommends retaining project records for at least seven years.

8 Using the specification

To reiterate the Preface, the specification (with the drawings) has nine functions:

- Design record.
- Demonstration of statutory compliance.
- Cost planning tool.
- Tendering document, both for contractors and for subcontractors.
- Contract document.
- On-site manual.
- Evidence in disputes.
- Facilities management resource.
- Project management.

The specification must be prepared with all these functions in mind but, because there is bound to be some conflict between them, and because it is usually assumed by the specifiers that only a single version is to be produced,[1] these functions must be prioritised.

For better or worse, the contractual function is the most important and must override the specifier's urge to provide a user-friendly document, where this might lead to conflict. This chapter examines the special needs of the various users, and suggests ways to address them without conflict.

Design record

The descriptive contract documents (specification and drawings) record design decisions progressively as they evolve. This is an automatic process, a simple by-product of their creation. Even so it is a useful outcome. Decisions on one project can save time on later projects for example, particularly if pre-edited versions of office or national masters are used, rather than the 'rolling specification' system. Especial care needs to be taken with decisions

1. It is possible for alternative versions to be readily produced from specifications text in a database, for example.

for such potentially influential documents as they will impact on many projects.

Standen (1995), perhaps overstating the case, argues that the specification is primarily a design document. Nevertheless, while other functions are more important, the specification is the only place that many design decisions are recorded.[2] It must be emphasised that only the final decision is recorded in the specification. Intermediate decisions are usually not recorded at all (this is not the case with drawings, which are developed through a series of standard iterations, each recording intermediate decisions). The exception is specifications for design and construct, which describe the design problem rather than the solution. For these, the contractor must prepare and submit response documentation, which will describe the solutions, for example in PC-1 (1998, clauses 6.2 and 6.3).

So, except for design and construct, the research, the rationale, the performance requirements, the reviews behind the decision are not recorded in the specification itself, and nor should they be. Separate files are necessary for this background information, which is especially important when substitutions are proposed or when specification content is queried, as it may be by the client, the contractor or the courts for example. The *NATSPEC Commentary* records the rationale behind many of the decisions recorded in its *Template* (default) material, for example. Specifiers are therefore in a better position to review such material, altering it if need be for a particular project.[3]

The specification records these decisions incontrovertibly. There is no backing away from it. Specifiers cannot say, "that isn't what I meant". It is important, then, that design intent is communicated clearly to the specifier in the first place, and that the designer reviews and checks the specifier's design decisions and interpretations.

On larger projects, many specification and detailing decisions are made without the principal designer's input. Few such designers have the interest or time or knowledge to be involved in resolving every detail of their projects. For architects, the need to check the specification applies especially to aesthetically important components of the building. These components may include services, such as handbasins, luminaires and exposed ductwork. This checking is vital where subcontract external specifiers are used –internal office communications e.g. between architectural designer and architectural specifier can be bad enough. To minimise problems of discontinuity, ideally both the designer and specifier – or at least, the project architect/engineer and specifier – should be the same person.

Lastly, the contract administrator can often only be advised (indirectly) of design intent via the specification and the drawings. Recording the decision incorrectly or imprecisely runs the risk of an unsatisfactory outcome. Review of the specification (and drawings) by all three parties – designer, specifier/documentor, and contract administrator – is a good idea if possible. Needless to say, the program must allow time for this, and it must be done before going to tender. Otherwise it will be too late to efficiently correct errors and ambiguities.

2. Different types of design decisions are recorded on the drawings. Specifying is clearly a design process, as is documentation generally.

3. The specification is actually a communications nexus. It is the link between the design process and the construction process but, for construct-only projects at least, contains no explicit information about either. One could, using the specification as the starting point, try to work backwards to determine design intent. Indeed, the contract administrator may attempt this in dealing with substitutions, in effect, reading between the lines. The contractor routinely uses the specification as a starting point for determining construction methods, sequence, sources and so on. In other words, the designers condense their requirements into a specification, and the contractor just adds water.

Statutory compliance

The compliance function of the specification begins with the development approval, where the embryonic specification takes the form of schedules of external materials and finishes. These schedules are expected to reflect any advice arising from consultations with the planning authorities, including aesthetic requirements e.g. for projects in heritage zones. These development approval schedules are in turn used at the design development, or early documentation, stage as a basis for the building approval or 'compliance' specification. The project is committed to them.

In New South Wales the *Environmental Planning and Assessment Act 1979*, Section 149C states that the council *may* require plans, specifications, survey reports and certificates to be supplied by the applicant. Similar, if not more explicit, requirements exist in most Australian states and territories. The purpose of these documents is to facilitate assessment by building control officers of the building proposals against the BCA and other legislation under the control of local councils. In some states this purpose is explicit in the legislation.[4] Similar documentation will need to be submitted to various other statutory authorities before the building can get underway. In particular, proposals for gas, electricity, and hydraulic services need approval. The relevant authorities and regulations for this work vary around the country, though hydraulic services are progressively coming under local authority control.

For BCA compliance, the requirements for the drawings are usually spelt out. Essentially they are the general arrangement drawings, largely comprising floor plans and elevations. Details are not required. This corresponds to maybe 20-30% of the drawing work required to build a commercial or institutional project, particularly given that the plans are not usually adequate for construction at this early stage.

The requirements for the compliance specification are not so clear. Is a complete specification i.e. one suitable for construction, required, or is an outline specification adequate? Probably the regulations generally anticipate something like a 20% complete specification, to match the compliance drawings. Indeed, relatively little in a full specification deals with regulation-oriented 'compliance' material in any case. The building control officer, for example, is not interested in the specification of veneers for the reception desk. So it is this 20% which needs to be submitted for approval – no more, no less.

However, stipulated requirements for the contents of the submitted specifications vary from state to state. Common requirements include

- description of materials;
- demonstration of compliance with building regulations;
- construction;
- second-hand materials;
- methods for drainage, sewerage and water supply;

4. The South Australian Development Regulations 1993 Schedule 5 requires the developer to provide such "information as may be necessary to show that the building work will, if performed in accordance with the specifications and drawings, comply with the Act and these regulations ...".

- methods for demolition and a description of the building being demolished;
- safety of the public; and, for alterations,
- the class and use of the existing building.

Of course, the compliance specification cannot be the complete contract specification. At the time of the submission of the building application, the project documentation is usually in its early stages. In fact, specifications are commonly prepared in earnest only at the tail end of the documentation program.

But, because of uncertainty as to what is required, applicants will often submit a pseudo-specification (i.e. a complete specification cobbled together quickly from previous specifications, with little project-specific material), on the assumption that a full specification may be required by the building control officer and should be submitted just in case. On very few projects, if any, would the 'full' specification submitted at this stage be the final or near-final project specification. The documentation program makes this simply impossible. Accordingly, the building control officer has to wade through a great deal of extraneous, and largely dubious, text to find the compliance material. Judging from the number of times such specifications seem to be queried by building control officers, it would appear that they are often simply not read at all, which rather defeats their regulatory purpose while perhaps suiting the applicants' purposes rather better.

The compliance specification could default to the deemed-to-satisfy solutions to performance requirements, such as AS 3700 (1998) (see below). If BCA defaults are to be exceeded, this must also be made clear.

This material could be supplemented by general project-specific information such as materials and finishes schedules, and all could be submitted as the basis for approval. Ideally, this specification would be organised to match the BCA, rather than traditional worksections, to make it easier for the specifier to cross-check during preparation and the building control officer during assessment. Consultant building code specialists, if any, would also find such a structure useful in their pre-building approval compliance check (Gelder 1995b).

With the shift towards performance-based building control legislation, the specification is becoming increasingly important as a means of demonstrating the proposed mode of compliance. For example, the BCA states, under Structural Provisions at clause BP1.1, that:

A building or structure including its materials and components must be capable of sustaining at an acceptable level of safety and serviceability –

(a) the most adverse combinations of loads ... ; and

(b) other actions, to which it may reasonably be subjected.

Under the deemed-to-satisfy provisions at clause B1.3(a), the BCA states that the requirements of BP1.1 are met for masonry construction if it complies with AS 3700. If the designers decide to run with this option, as they probably do in 99% of cases, then this must be stated in the documents they submit to the local council. If they decide to do something else, then they must spell this out in rather more detail.

The project-specific requirement to comply with AS 3700 must also be passed on to the builder. The specification is the appropriate vehicle for both these tasks. This is especially important where the builder has to both design and construct the masonry. Building control officers are interested in both aspects. One way of looking at the compliance specification is as the second part of a two-part compliance document, the first part being the BCA.[5]

Contractors will not use deemed-to-satisfy standards such as AS 3700 (other than voluntarily) unless they have been specified. This fact is often overlooked by those who prepare such standards, and by any specifiers who think that such standards are mandatory and so do not need to be specified. This applies to all the deemed-to-satisfy text in the BCA, such as clauses C3.15 and C3.16 for fire-stopping, which in turn reference several standards, and BCA Specification C3.15 (particularly Part 7). Accordingly specifications, instead of relying on prescriptive BCA requirements to apply come what may, must now cite these requirements if they are to apply. Such citations are likely to be extensive.

The building control officer is the compliance specification's main target. The specifier is the other principal user, assuming that the project proceeds beyond the building approval stage. Many clients authorise ongoing documentation pending building approval, in any case. The specifier will use the compliance specification document in two ways:

- As a check for compliance of the contract specification, at various stages. The project specification must match the approved compliance specification, altered as required in the building approval conditions.
- As a starting point for preparation of the contract specification.

For both these purposes the compliance specification may need to be restructured from the BCA-aligned version submitted to council (if this format is used), to the construction-oriented version used contractually. Use of a database to prepare the specification would be a great help, because the appropriate report format could be used to call up either the compliance version or the contract version.

Building control officers need to insist that all that is required – and no more – is submitted. Specifiers would not waste time cobbling together a 'full' pseudo-specification and building control officers would not have to waste time wading through it. Everyone would benefit.

Use of national building master specifications, which can be closely aligned to the national building codes (where they exist) is a good idea.[6] Unfortunately, building services regulations may vary from jurisdiction to jurisdiction. [7] Thankfully, however, the basis for the requirements remains fairly constant nationally e.g. the Australian standards AS/NZS 3000 (2000), and AS/NZS 3500 (various).

5. *Actually, it is a three-part document. State and territory variations to the national code comprise the second part.*

6. *For example, NATSPEC is aligned to the latest BCA Amendment.*

7. *See for example the New South Wales Code of Practice – Plumbing and Drainage 1992.*

Cost planning

The cost planner for the project, usually a quantity surveyor, will do a series of cost checks during documentation, culminating in the pre-tender estimate. If the specification proceeds in parallel with the drawings, the quantity surveyor will be able to use both documents alongside each other. This makes the estimates more precise than if only the drawings are provided, or if the specification has not been developed in this way e.g. the structural drawings are well underway, but the structural specification has not started, or vice versa.

For successive cost checks, it would be useful to flag changes in the specification from the previous version, as is usual for drawings. For estimators, precise cross-referencing between the specification and drawings is very useful. For precise estimating, the less subjective the documentation the better. Generally, specifiers should follow the principles already outlined. If the estimator cannot cost the specification accurately e.g. if everything is "to approval", neither can the tenderers, making the pre-tender estimate and the tenders themselves much less reliable.

One side benefit to specifiers of progressive cost checks is that the quantity surveyor will query unresolved items, conflicts and so on. This second pair of eyes is very useful indeed.

In terms of programming, quantity surveyors would often rather, at the 50% stage, have 50% of the specification sections 100% complete, than 100% of the sections 50% complete. The same applies to the drawings. Lots of just-started 'blue-line'[8] drawings are of little or no use in estimating. Check whether this is the preferred approach with the cost planner on each project.

Tendering document

Tendering is often a hurried process, with tenderers looking for ways to speed it up without too much risk of error. Contract tenderers will usually need to break the documents down into smaller packages for pricing by subcontract tenderers. With tendering, documents have to be analysed to generate quantities (if no bills are provided), to come up with proposals e.g. for performance-specified work and substitutions, and to price it all. If this proves too difficult, short cuts will be taken.

Specifiers can help. First, the document can be loose-leaf and single-sided, to help with photocopying; it could even be provided on disk. Second, worksections should be as small as practicable, with self-contained subsections and clauses. If worksections are small, further disassembly is unlikely to be needed. If subsections etc. are self-contained, any required disassembly will be painless. Unfortunately, not all NATSPEC material is structured like this particularly in terms of self-contained subsections and clauses. As stated in Chapter 5, specifiers are unlikely to be able to anticipate the subcontract arrangements if only because different tenderers will have different approaches.

Third, non-standard requirements could be separately identified. Use of a two-part approach to the specification (e.g. relying on *NATSPEC Reference*) will help with this.

8. Preliminary drawings were often set out in blue pencil, which does not print.

Everything in the second part, the project specification, is non-standard, and everything in the first part is – or should be – standard.[9] Similarly, advice in *Tendering* as to which performance-specified work must be responded to with a proposal at tender, and which proposals can be delayed, will simplify matters. This will also apply to proposed substitutions, selections from lists of alternatives and so on.

Major substitutions (mostly against proprietary items) may need to be proposed at tender, and tenderers will often submit non-conforming (and cheaper proposals) to try to win the job. The bulk of the building specification should be in more open terms so that, to get value for money for the principal, the contractor is able to compete at tender by shopping around, at least for major items such as concrete and airconditioning.

Fourth, the specification should cross-reference the drawings with precision, rather than using "as detailed". This helps the contract tenderer locate details relevant to each subcontract tenderer. The converse, drawings precisely cross-referencing the specification, is also a good idea.

If a bill of quantities is provided,[10] specifiers should be aware that tenderers (especially subcontract tenderers) may well price off this document without reference to the specification, given pressure of time. Cross-references from the bill to the specification will encourage subcontract tenderers to look at both (and to ask the contract tenderer for a copy of the specification). Specified bills are another approach to solving this problem.

Finally, tenderers should be given enough time to do a proper job of pricing. Recommendations from the Construction Industry Cooperative Committee of Minnesota are given in Table 8.1. Extra time should be allowed for prior approval of products or design proposals. It goes without saying that the documents must be available to tenderers from day one of the tender period.

Contract document

The specification, as part of the contract, provides the written part of the description of what the contractor is to provide. As such, it represents the principal's needs and desires. The principal will expect advice from the specifier, and may well leave the whole business entirely in the specifier's hands, but the specifier must at all times act in the principal's best interests.

Estimated cost	Simple project	Average project	Difficult project
To $US1 million	24	28	32
$US1 to $US5 million	30	32	36
Over $US5 million	36	38	42

Table 8.1: Minimum estimating times in days (Wolfe 1994)

9. Two-part documentation is widespread in the construction industry, because of the many advantages it offers. Examples include the BCA (see also sidenote 5), standard forms of contract (appendices are the second part), standards (the specification is the second part, except for ISO/IEC standards adopted in Australia where the second part is the appendix describing Australian variations) and master specifications systems. The second part should cite the first unambiguously, stipulate precedence, not repeat material in the first part, and explicitly state which bits of the first part either apply unchanged or do not apply, are modified and how, and/or are supplemented with new text provided (Gelder 1995c). CIS trialled the two-part approach to specifying, providing the first part in book form (NATSPEC Reference) and the second part (Schedules) electronically and in loose-leaf form, from 1994 to 1997.

10. One of the advantages of bills is that they expedite the tendering process. Tenderers do not need to work out quantities themselves. On the other hand, provision of bills by the principal tends to split the responsibility for tendering, which is at odds with Abrahamson risk allocation principles.

This means, or should mean, that what is specified is important, and so

- is not to be ignored by either the contract administrator or contractor or tenderers; and
- is to be followed through on-site, by both the contract administrator and the contractor.

Too often though the specification is ignored, and is not followed through. Like the drawings, the specification is not a document of last resort. It is intended for everyday use in the execution of the contract.

For the specification to be used in its contractual sense, it must be meaningful, contractually sound, technically correct, and editorially resolved, otherwise it invites frustrated users to relegate it to the bottom drawer. Most of this book is about how to satisfy these requirements. And, as said before, the contractual (legally binding) role of the document must rule the content and style of the specification. The other considerations outlined in this chapter are therefore secondary. The urge to assist users, particularly in terms of repetition, must be resisted where it conflicts with contractual requirements. Where there is no conflict, by all means help them.

On-site manual

On-site, the two main parties using the specification are the contractor and the contract administrator. If the specification is well written, both will interpret it the same way. Strictly speaking, neither party is supposed to read between the lines. Another set of entities, the manufacturers and suppliers, will also be using the specification, and are particularly interested in the substitutions process.

The contractor's needs regarding specification structure are similar to those of the tenderers (discussed above), and then some. The contractor's operations are split into head office activities, and on-site activities. Generally, head office orders the products (manufactured and fabricated) needed on-site,[11] and on-site operatives assemble these into built products. Ideally the written description should be split along these lines also. For example, for ordering purposes, head office will require quantities (not in the specification), and quality e.g. brand, grade, finish. Information on the quality of manufactured and fabricated products is in the specification.

On-site, operatives will take whatever is delivered, assuming it to be what is specified (quality) and of the quantity required. 'Just-in-time' delivery is preferred, to minimise the risk of damage or theft in storage but this leaves little time for careful checking against the specification on-site. These operatives are interested in assembly (best shown on the drawings, including relevant annotation), and quality of execution (the built product – shown in the specification). For traditional trades, such as brickwork, they may well assume that the specification has nothing useful to tell them (a case of teaching grandmother to suck eggs) but as even traditional technologies are changing rapidly these days, they will

11. On very small contracts (or subcontracts) this is done on-site using the ubiquitous mobile phone – a sort of mobile head office (with the ute) – sometimes by a visiting, rather than resident, contracts manager. On very large projects a project office is often established, at or near to the site.

probably be wrong in this assumption. Indeed, a survey (Nawar and Zourtos 1994) found that bricklayers were not familiar with the masonry code (AS 3700) and that their work did not, therefore, comply in many respects e.g. wall tie spacings.

The difficulty, then, is that they will look at the drawings, but they will not look at the specification. Even when they do, if it references a standard they will almost certainly go no further since the standard will not be held on-site. Remedies for this include follow through on-site by the contract administrator, against the workmanship standard, and the ready availability of easy-to-read workmanship standards, perhaps on a CD held on-site, or in a book of on-site execution extracts from construction standards. Proper ongoing training of tradespeople, against the standards, is also essential.

It is clear that the specification can legitimately split materials (manufactured products) and execution (fabricated and built products), particularly since the materials may well be the subject of supply-only agreements. This is well handled by the North American SectionFormat approach. NATSPEC's system is not dissimilar, but is a little more sophisticated (see Chapter 4).

Product selection by the contractor, maybe involving substitution (often unauthorised),[12] is another specification issue at this stage. The contractor will seek to maximise its own profit during the contract by shopping around for products and suppliers – some assumed at tender may be dropped. Contractor considerations in product selection can be quite different to those considered by the specifier. They are often process-oriented, and include such things as time saving, capital cost saving, advice on installation, ease of erection, returns e.g. unused paint, and other environmental or OH&S issues such as whether or not the product is safe to install without protective clothing.

However, contracts may oblige contractors to select products "which [are] both suitable for [their] purpose and consistent with the nature and character of that part of the Works for which [they] are used" (JCC clause, 1994a, 6.08.01). It is a fallacy to think that if the specification is wide open the contractor can use the cheapest material available (also, Billingham 1995).

If contract administrators, the second main party using the specification, come from the same practice as the specifier and/or designer, they tend to read between the lines. They may well have a very clear idea of design intent, particularly if the practice has a strong design tradition. Specifiers often assume this familiarity, hence the widespread use of "to approval". This relies on the contract administrator understanding the design intent which, as stated earlier, is not given in the specification. If the contract administrator is completely independent, the specification has to be more robust, more self-contained – written, in fact, as it should be no matter who the contract administrator is. The contractor, after all, does not have the benefit of the contract administrator's inside knowledge, and so cannot price the work "to approval"!

12. The Barbour Report 1994 (Billingham 1994) suggests that half of all UK contractors admit to not always knowing when a subcontractor has made a change, and that 40% of them broke specifications themselves. The principal can compel the contractor, in theory, to rectify the works to accord with the original specification, but this is usually so disruptive that it is not done. A good example of this is the UK case of Ruxley Electronics & Construction v. Forsyth, in which the House of Lords rejected a claim by a principal for the cost of rebuilding a swimming pool in accordance with the specification, since the reinstatement cost would be out of proportion to the loss of amenity (Westgate 1995).

From time to time the construction industry, or parts of it at least, argue that the contract administrator should be completely independent of the documentation team. Certainly the contract administrator, in the conventional scheme of things, is in the difficult position of having to wear two hats, that of quasi-arbiter, and that of principal's agent. A sound specification, where the contract administrator does not have to read between the lines, is a very useful defence against charges of bias, in that whoever the contract administrator is, the interpretation will be the same as the contractor's.

One of the contract administrator's functions is to ensure that what is built conforms to what is specified. To do this, the contract administrator must regularly 'observe' the works in progress. What this entails depends partly on the contract. For example, AS 4000 (1997) clause 39.2(a)(iv) says that failure by the contractor to use the materials or standards of work required is a "substantial breach". This is not qualified, so appears to apply to any such failure – even to the use of one non-specified material in one case somewhere on the project – in principle.

In practice, the contract administrator is unlikely to be so tough. A substantial breach requires written notice to the contractor to show cause why either the work should not be taken out of the contractor's hands or the contract should not be terminated. If no cause is given, the principal can do either. So the consequences are arduous. In any case the contract administrator, though required to watch for failures and no matter how assiduous, would almost certainly not pick up trivial failures.

Just what sort of non-compliance the contract administrator is expected to spot is the subject of debate. As discussed in Chapter 5, the RAIA is keen to show that the contract administrator is not expected to examine every nail for compliance (seeming to take the argument too far, however, by disputing the need for any verification against cited product standards), and this is correct. But, if the quality of nails has been specified, then the contract administrator might reasonably be expected to check compliance in some way – otherwise, why was it specified?

It is suggested that lessons be drawn from third-party product certification (discussed in Chapter 2). One technique is a formal approach to product sampling and statistical extrapolation. This depends on sample size and how much of the particular product can be found in the project (for which a bill of quantities would be useful). If there is only one instance, then the contract administrator has no choice but to check it for compliance. If there are two instances, then only one need be checked to gain 50% confidence in the general compliance or otherwise. If there are 100 instances, then perhaps 10 would be checked, giving 10% confidence (nails would be checked by the box, not individually, by the way). And so on. Sampling frequency is specified in several NATSPEC worksections e.g. *Concrete post-tensioning*, under Tests.

What sort of confidence level is required? This depends on what is acceptable to the principal, which could be stated in the client/architect agreement. The higher the level of confidence required, the higher the fee for this part of the contract administrator's work. The confidence level might also reflect how confident the contract administrator is that the contractor can deliver the specified quality. This, in turn, might depend on whether or not the contractor is quality assured. Confidence also depends on the criticality of the product. Products providing for health and safety e.g. structural products, will generally be more important than others, for example.

The means of verification must be clear, and will vary depending on the specification method e.g. for performance specifying, verification will be through testing whereas for specifying by reference to a standard, verification may be through third-party certification to that standard e.g. using the StandardsMark. A number of these verification procedures will be hold points (the work in question cannot proceed until the verification procedure has been carried out successfully); others will be witness points (the contract administrator may, or may not, observe the work in question at that point). The specification needs to make this clear, too.

Quality assurance minimises the number of hold and witness points involving the contract administrator[13] because many of these points will be internal to the contractor and delineated in the contractor's project quality (inspection and test) plan, which the auditor, but not the contract administrator, reviews. On such projects, then, the contract administrator checks products, and the auditor checks processes. On finding a non-conforming product, the contract administrator may query the process that led to it, asking, for instance, why it was signed off under the contractor's inspection and test plan.[14]

On normal (quality control, so called) projects, no-one external to the contractor checks processes, and the contract administrator checks many more products and at many more points in the process. Under Abrahamson, the contract administrator cannot exert quality control. Control is in the hands of the contractor, and so the contractor must be responsible for the quality of execution being as specified. This seems like a good idea, but to succeed, the specification must, again, be robust, and self-contained, with quality spelt out explicitly, and quantitatively. Specifiers have not yet come to grips with this requirement.

Another function of the contract administrator is the issue of variations to the contract. These arise for all sorts of reasons, only some of which the documentation team can control. A US survey of variations (change orders) analysed the causes, including 349 change orders that arose because of errors and omissions in documentation. Results are summarised in Table 8.2 (Preston 1999). Five causes accounted for 71% of change orders, echoing the Pareto principle. Maybe 50% of the overall total are attributable to the architects.

13. See letter from John Pilley (1996) regarding excessive use of hold points under QA.

14. This interest in processes from the principal's side of the contract is virtually precluded in conventional contracts, either implicitly or explicitly, as discussed. They are none of the specifier's business, hence the problems with discontinuity and Cornick's (1991) views on quality assurance. Contracts specifically geared to quality assurance seem to be warranted.

Manufacturers are naturally keen to have their products used in projects. Getting them specified is only part of the story (see Chapter 6). For the many products not specified in proprietary terms, the brand has yet to be selected, and for those that are specified in proprietary terms, substitution is always possible (Gelder 1997c).

Manufacturers will generally be dealing with subcontract suppliers and installers, who are usually left to their own devices to sort out the brands of products used,[15] again often unauthorised.[16] Contractors therefore do not usually get involved in brand selection of small or simple items such as nails or plasterboard. Subcontractors find it difficult to challenge any selections or deselections (and any requests for them) made by contractors, because of a fundamental power imbalance. Some contractor selections or deselections are made for expedience, with a view to the overall program, and the contractor may be quite inflexible on these. With this in mind, the contractor may be prepared to listen to a proposal that could save more time.

One problem for subcontractors is that they do not always get all the relevant project information. It can be difficult for the contractor to rearrange the specification and drawings for dissemination to subcontractors. This means that, if the subcontractors have not got all relevant information, then neither will their material suppliers. Manufacturers may be missing a vital piece of data when asked to quote.

Error or omission in documentation	% of 349 change orders
Architectural documents	20
Coordination of drawings and specifications	19
Site investigation – verifiable conditions not considered	13
Electrical documents	10
Mechanical documents	9
Other reasons • Survey – conditions not documented or accounted for • Record documents – conditions not documented or accounted for • Program (i.e. brief) communication – communicated requirements not included • Civil documents • Structural documents • Plumbing documents • Contracts and preliminaries • Phasing – impractical work sequences • Authorities' requirements – changes not involving codes • Codes	29

Table 8.2: Cause of variations

15. Research indicates that mechanical and electrical subcontractors select over 25% of products and 42% under design and construct (Coomber 1994).

16. A 1994 survey (anon 1994) revealed that 25% of UK subcontractors admitted to breaking specifications. But who is to blame? Consider these quotes: "[architects] have to have the courage to draw up a strong specification, identify clearly why they want it and them stick to it. If someone comes in with a cheaper but supposedly equal product, they should say: 'We'll give you a chance next time'" (quoting the manager of a cladding subcontractor), and "Some trades feel that by not standing by their specifications – or not bothering with them at all – architects are to blame for lowering building standards" (Coomber 1994).

Assuming that manufacturers are at the bottom of the chain of contracts, the manufacturer may now make proposals for substitution against the product selection. But this selection may have been made anywhere up the chain and it can be hard for the manufacturer to work out where and why. So while some substitutions can be resolved quickly, others may have a long way to go up the chain e.g. to the client, for a decision.

Manufacturers should take an interest in the intended use of their products. It may be that in spite of all the decisions made higher up the chain, the selected product is not suitable for the application, and that another in the manufacturer's range would be better. It may even be that there is no suitable product in the manufacturer's range, in which case another's product might be suggested. This interest is not altruistic. Manufacturers may be found responsible for defective products and an unsuitable product is more likely to become defective more quickly, and more expensively. Product selectors will also remember the product failure, even if the manufacturer was exonerated (Franklin 1999).

Evidence in disputes

Specification disputes with the contractor range from minor disagreements on interpretation, which are usually resolved quickly through arbitration, to full-blown court cases. Disputes may be in contract, or in tort. In all these, the specification can be a two-edged sword. If well written, it will provide an invaluable defence, but if badly written it can be used offensively by the contractor. Shoddy specifications are the contractor's friend. Since specifications are written to define the principal's requirements, such documents have clearly failed to perform. The findings of an American survey of nearly 200 construction situations that led to claims, litigation or arbitration, involving specifications, are summarised in Table 8.3 (Nielsen and Nielsen 1981).

Problem area	Percentage of total cases in which problem area was the primary cause
"Or equal" specifications	24.7
Ambiguous phrasing	12.5
Conflicts between plans and specifications	12.4
Defective specifications (i.e. buildability)	12.3
Inaccurate technical data	11.8
Product performance deficiencies	8.2
Inspection requirements	5.5
Health and safety requirements	4.9
Construction tolerances	3.1
Typographical errors	1.5
Other	3.1
Total problem areas	100.0

Table 8.3: Primary specification problem areas

Nowadays in Australia the percentages may well be rather different. For example, "or equal" specifications rate highly in this table largely because of US federal government contracting practice (see Chapter 6). Likewise, the issue of responsibility for health and safety depends on the legislative position.[17] In some jurisdictions the specifier and/or designer is responsible for a number of these issues, but generally common law makes the employer responsible for the health and safety of its employees. On building sites, the employer is the contractor. This responsibility is explicit in a few contracts, hence the conditions of contract requirements for workers compensation and employers' liability insurance, and liability for injury to persons. Any conflict on this matter, between legislation and contract, must be resolved. The problem in specifications arises when

- the specification requires use of a hazardous product or process, if the principal is responsible for health and safety; or
- the specification contains a requirement for health and safety (e.g. "Handle synthetic mineral fibres in a safe manner") if the principal is not responsible for health and safety.[18]

Specifications defective in terms of buildability are an interesting issue. In the United States, the Spearin Doctrine holds that the principal warrants, by implication, the sufficiency of the specifications supplied by the principal to the contractor, even though the principal did not prepare the specification. On the other hand, the specifier does not impliedly warrant the sufficiency of the specifications. The specifier's performance instead is measured in terms of negligence, rather than error. By contrast, much European legislation relies on strict no-fault liability i.e. fault does not have to be proven. If a building collapses, for example, the architect is automatically liable – guilty until proven innocent.

In Britain and Australia "the proprietor gives no implied warranty that the work can be done in accordance with the design prepared by him or on his behalf" (Bennett 1980).[19] Likewise the specifier makes no such warranty. For example, under the RAIA client/architect agreement, the architect does not warrant that the design and the documentation will be "fit for purpose".

However, it is not unknown for principals to demand an "express and unqualified warranty" from the designer and/or specifier to the effect that the documents are fit for purpose in every respect e.g. buildability, functionality, safety. To offer such a warranty is to risk voiding one's professional indemnity insurance. The RAIA recommends against such warranties, as they defeat the assessment of any errors in terms of negligence (see RAIA Advisory Notes AN10.01.603 [1997b] and AN08.02.606 [1997a], and Nelson 1991).[20]

Many other issues relevant to disputes, particularly ambiguity and conflict, have been discussed throughout this book. If you follow the guidelines presented here, the risk of dispute will be greatly diminished. To quote former RIBA President Max Hutchinson:

If God is in the details, then one part of the Trinity is in the specification. Have faith, specify with care, and hold firm (Hutchinson 1993).

17. See sidenote 1 to Chapter 2.

18. See RAIA Advisory Notes AN16.01.601 (1984) and AN10.01.610 (1992), which seem to be at odds with RAIA Advisory Note AN14.01.612 (1998d). In 2000, the RAIA's National Practice Committee agreed to establish a working group to comment on Worksafe Australia's Construction Hazard Assessment and Implication Review Guidelines, and to formulate a resolution to National Council regarding the importance of addressing OHS at design and the role that architects might have in contributing to this (anon 2000).

19. See also Dorter and Sharkey 1981. Things seemed to have shifted a little since, such that "...although there is high authority that the principal ... does not give an implied warranty for "buildability", there are other authorities coming close to it ..." (Dorter and Sharkey 1995).

20. In the contractual sense, one warrants one's own work, but guarantees the work of another.

Facilities management

Facilities management is a broad (and getting broader) approach to the management of buildings and other facilities in-use, all the way through from at least as early as commissioning to disposal. It deals with management of spaces, internal and external environments, building information including evaluations and audits, support services, and strategic issues e.g. when to refurbish, when to sell, when to lease.

The specification (with the drawings) is potentially useful to facilities managers, addressing facilities management issues explicitly in terms of the following:

- Commissioning and completion testing e.g. of services installations.
- Provision of operation and maintenance manuals (see RAIA AN17.01.100 [1997c]).
- Record drawings.
- Warranties.
- Training.
- Maintenance records and tools and spares.
- Maintenance itself, at least during the defects liability period.[21]

In certain circumstances, requirements for ongoing maintenance work may also be described, if the contractor is to have that responsibility. All these requirements should be isolated from the rest of the worksection, just as maintenance is separated in standards and the BCA.

The specification is also of use in a general sense, in that it defines what should have been built in terms of proprietary items, for example. This will help with maintenance, fit-out work, alterations and additions, and demolition. However, just as record drawings are required especially for services, so a 'record specification' should also be provided. This would record the actual proprietary items used in the project, and their sources. This would take care of substitutions (proprietary items specified may not have been provided), and contractor-selections against performance, reference or descriptive specifications – and where the specification is silent.[22] It is best prepared by the contractor and reviewed by the contract administrator.[23]

Ideally, the record and/or contract specification would be provided electronically, so that it can be searched easily, and be manipulated by the facilities manager to record changes-in-use. That is, along with drawings provided digitally, it could form the core of the building's electronic description or model. If the data were connected to attributes of the products used e.g. behaviour in fire, thermal performance, and recycling potential, the database could also be the basis for simulations of the building-in-use, and for intelligent building management systems e.g. controlling airconditioning systems, lighting, alarms and so on. Realising this simulations role of the specification (also extremely useful in the design phase) lies in the not-so-distant future.

21. See NATSPEC General requirements – Completion, for example, and RAIA Advisory Notes AN14.02.100 (1993b) – warranties.

22. Such a document would also be very helpful in disputes.

23. Alternatively, the built fabric itself could provide this information, through use of bar coding. So-called 2D bar codes can hold several A4 pages of information (e.g. sources of spares, maintenance information, Material Safety Data Sheets [MSDS]), and can be printed or etched directly onto building products (anon 1996). This would assist contract administrators, as well. If required, they should be specified.

Project management

The specification is also appropriate for informing contractors of many project-specific requirements for a number of management aspects of the construction project, quite apart from the obvious issue of quality. Likewise, the drawings and general conditions, as well as other contractual and non-contractual documents, touch on management issues.

Here the specification is examined from the point of view of the nine project management functions – integration, scope, time, cost, quality, risk, procurement, communications and human resources – to show how these functions are, or can be, dealt with.[24]

Conversely project management considerations influence the documents themselves. The contract documents as a whole must be clear, concise and logical (communications) and fair (risk). They should not contain material that does not apply (scope). They should be prepared in a timely manner (time) and should set appropriate quality levels (quality). Buildability should be considered (cost, time). Procurement considerations impact on methods of specification e.g. by proprietary item. Human resources management is perhaps a minor influence on the documents.

All contract documents exist as a crucial part of risk mitigation, scope definition and project communication, for the benefit of all parties. A bit of extra effort in documentation, no matter how small, will pay off throughout the project, due to the magnification effect of documentation activities (relatively small effort) versus site activities (large effort).

Table 8.4 gives a summary matrix of documents versus project management functions.

	Project management functions								
Documents	Integration	Scope	Time	Cost	Quality	Risk	Procurement	Communications	Human resources
Tender documents	●	●	●	●	●	●	●●	●	●
Conditions of contract	●●	●●	●	●	●	●●	●●	●●	●
Preliminaries	●●	●	●●	●	●●	●	●	●	●
Technical specifications	●	●	●	●	●●	●		●	●
Drawings	●●	●●		●	●	●		●●	
Bills of quantities	●●	●		●●	●	●		●	
Statutory requirements					●●	●			●●

● – *link between document and project management function*

●● – *significant link between document and project management function*

Table 8.4: Documents v. project management functions

24. The United States Project Management Institute's (PMI) nine Project Management Knowledge Areas (as described in PMI Standards Committee [PMI 1996]) were used by the Australian Institute of Project Management (AIPM) in its formulation of the Australian Industry Standards for Project Management.

Documents provided to tenderers may include all or some of the documents above. They directly affect, or reflect, procurement. For example, if no documents are provided then presumably the contract is based on outright nomination. This is risky and is suitable only in very rare circumstances e.g. small works, complete trust in nominee, specialist or unique skills required and time critical. One possibility is the provision of an artwork e.g. a sculpture.

More common is the opposite extreme, the provision of complete documentation comprising conditions of contract, drawings, specification and supplementary material such as reports, bills of quantities and scope (interfacing) statements. These documents are expected to be final and fully coordinated before going to tender, because the principal takes all the risk for information provided – the more provided, the higher the risk. The principal needs to control the design and documentation carefully to minimise this risk. This 'traditional' approach implies that the project is larger or more complex, that tenderers are not fully trusted (and so antagonism is almost inevitable in the construction phase), the work is knowable and conventional, and time is not critical.

A middle ground is becoming more common, where the contractor provides a combination of design, documentation and construction. Clearly less than complete conventional documentation is required. If design is part of the package, the responsibility of the contractor is substantially increased, including a responsibility for 'fitness for purpose', for which the principal otherwise has no recourse (see Department of Defence 1992). Use of these hybrid approaches implies that time is critical, risk to the principal must be minimised, the tenderers are trusted (though quality assurance may be substituted for quality control), and that the projects are large or complex. It may also be that much of the work is not knowable, or is innovative.

Lastly, tendering on incomplete documentation, such as building approval drawings and outline specifications, is widespread for industry sectors such as residential (alterations and additions). Detailing is left to the contractor and relies on 'traditional' construction methods being used. Use of this approach, where these methods are not being used, may carry undue risk. Low profit margins for designers working in these sectors, on the other hand, virtually dictate this approach.

Other characteristics of the tendering process, which carry project management ramifications, include method (negotiation, selected tenders, open tenders), contracting arrangements (single-prime, nominated subcontracts, multiple-prime) and ethical tendering (see *No Dispute* [NPWC/NBCC 1990] and AS 4120 [1994]).

The *Preliminaries* worksection of the specification, addressed to the head contractor, touches on all project management functions. Prime-contract preliminaries should be compatible with those used for subcontracts, just as the prime-contract conditions of contract should be compatible with the subcontract conditions.

Table 8.5 provides a schedule of NATSPEC *Preliminaries* subclauses from NATSPEC

Preliminaries subclauses	Project management body of knowledge functions								
	Integration	Scope	Time	Cost	Quality	Risk	Procurement	Communications	Human resources
General conditions	●					●			
Interpretations	●							●	
Site restrictions						●			●
Occupied premises	●					●			●
Protection of persons and property						●			●
Existing services		●						●	
Adjoining property	●					●		●	
Proprietor's site office									●
Protective clothing						●			●
Temporary services		●		●					
Use of existing services		●		●					
Temporary fence		●				●			
Provisional sums	●	●		●			●		
Order of work			●				●		
Program of work	●		●						
Site meetings								●	●
Requirements of authorities	●	●	●	●		●	●	●	
Safety	●					●		●	
Surveys					●				
Survey marks					●				
Items supplied by proprietor	●	●					●		
Changes to existing								●	
Progressive cleaning					●				
Run off						●			
Anticipated progress claims			●	●					
Progess claims breakdown				●					
Tax exemption				●					
Method of measurement	●	●		●					
Removal of plant		●							
Restoration		●							
Pest eradication		●							

Table 8.5: Importance of the *Preliminaries* as a project management document

BASIC, aligned to SBW-2, versus the nine project management functions, illustrating the importance of the *Preliminaries* as a project management document.

The technical specifications within a full specification primarily define quality but will also impact on cost, time, scope and risk. They often respond to options and defaults in referenced standards, many of which are cited in the BCA or in other statutory instruments.

For small projects, or in some disciplines, little or no technical specification is provided. It is replaced by notes and schedules on the drawings. Another easy option, blanket reference to default documents, is sometimes made with no qualification. The specifier relies entirely on context and the drawings. Again, many technical specifications are regurgitated specifications from previous projects. These are dangerous in terms of inclusion of irrelevant or inappropriate material, and exclusion of essential material.

For communication, this is all very unhelpful. It is also ambiguous in terms of scope and quality, increasing the principal's exposure to risk. Specifications, like drawings, need to be carefully prepared for each project.

Self-contained (one-part) specifications are a traditional solution, for architects at least. Such documents, whether prepared from NATSPEC material or not, need to be read carefully from cover to cover by all users, to locate critical items (project-specific rather than general). They must be carefully edited to avoid conflict between, say, reference specifications and proprietary specifications. Again, for communication they are poor performers, and so raise the risk of errors, conflict and ambiguity.

A recommended route for specifications is the two-part specification, referring to a reference specification (e.g. *NATSPEC Reference*) on the one hand, and providing a project specification (comprising all project-specific requirements and nothing else) on the other. Once the user is familiar with the reference document (quick for specialists) only the second will need to be referred to in most cases. Use will be quick and easy and conflicts or errors obvious. The principal's risk should be reduced, and communications enhanced.[25]

Other characteristics of technical specifications include method of specification (performance, proprietary, descriptive, construction method), contract administrator interventions (approvals, hold and witness points for inspections and tests, samples approvals, quality assurance), plain language drafting and other aspects of drafting, including cross-referencing, indexing, navigation, format and the like.

All these impact on the project management functions. For example, specification by proprietary item means that the consultants have assessed the item's fitness for purpose and accept (some) responsibility for it. Insistence on approval of minor items unnecessarily exposes the superintendent to risk. *No Dispute* (NPWC/NBCC 1990) advocates the transfer of responsibility to those best able to control the risk. Lastly, good drafting of specification clauses improves communications.

25. It has already been mentioned that two-part specifying has not proved popular with specifiers, who are perhaps not prepared to experiment on their projects with such a different approach.

Design consultants, wittingly or otherwise, play a major role in project management through contract documentation. Principals and project managers must take both the conditions of contract and the specification (particularly the *Preliminaries*) seriously, and should be closely involved in reviewing the content of the project *Preliminaries* with the documentation consultants, as an adjunct to the selection of appropriate conditions of contract. Conversely, documentation consultants should be familiar with and promote the project management role of the specification and other documents.

The quantity of documentation, and the quality level stipulated, should vary according to project life (other less noble factors, such as fees, may also come into play). It could be argued, however, that the quality of documentation prepared by professionals should be consistently good and not fair, or even poor, as it often is at present.

Appendix 1 — Historical specifications

This Appendix supplements the general historical material presented in Chapter 1. A sample of a complete contemporary brick and block construction master specification, based on NATSPEC material, is included in Appendix 4.

As far as possible translations are by the author. Untranslated words are in roman text. Words inserted for sense are in round brackets. Explanatory material is in square brackets. Gaps in the extract or in the original are indicated by ellipsis.

Court of the Tabernacle, Sinai, Egypt (950-500 BC)

The bible offers several specifications, including those for Noah's Ark (see Chapter 6), and a set given to Moses (considered a contemporary of the pharaoh Merneptah, c 1220 BC) for the Ark of the Covenant, dealing with materials supply (by donation), the ark itself, a table, a lampstand, the tabernacle and tent made of hangings, planks and bars for the tabernacle, a veil, an altar, the court of the tabernacle, the lamp, and the priestly vestments.

These biblical specifications may well have been very influential, probably in an unselfconscious way, in many cultures (Jewish, Christian and Muslim) over the centuries. They would have been read by building owners, designers, specifiers and contractors for example. According to Kent (1985), the design of the tabernacle itself was an important historical source for the medieval architect. It was studied by Renaissance architects such as Villapandi, and has influenced the work of Le Corbusier and Quinlan Terry (who has suggested that the tabernacle is the source of the Greek Doric and Ionic orders). It is therefore conceivable that the descriptive techniques used may have served as a parallel model for project specifications.

This specification is extracted from the biblical description in Exodus 27. There is enough information here to reconstruct the court – there were no drawings (see Kent 1985). Nor was there a contract! Though the tabernacle and so on are presumed to have been made c. 1200 BC, the first five books of the bible (including Exodus) were written by four principle authors or groups of authors over the period 950-500 BC, long after the events described (Grant 1984). As Grant says, "The surviving descriptions of the Israelite Tent [or, tabernacle] are much later, and include anachronistic elaborations; but they go back to very ancient times." The Hebrew cubit was about 500 mm (Dilke 1987).

The Lord spoke to Moses and said: ... Make the court of the Tabernacle. For the one side, the south side facing southwards, the court shall have hangings of finely woven linen a hundred cubits long, with twenty posts and twenty sockets of bronze; the hooks and bands on the posts shall be of silver. Similarly all along the north side there shall be hangings a hundred cubits long, with twenty posts and twenty sockets of bronze; the hooks and bands on the posts shall be of silver. For the breadth of the court, on the west side, there shall be hangings fifty cubits long, with ten posts and ten sockets. On the east side, towards the sunrise, which was fifty cubits, hangings shall extend fifteen cubits from one corner, with three posts and three sockets, and hangings shall extend fifteen cubits from the other corner, with three posts and three sockets. At the gateway of the court, there shall be a screen twenty cubits long of finely woven linen embroidered with violet, purple, and scarlet, with four posts and four sockets. The posts all round the court shall have bands of silver, with hooks of silver, and sockets of bronze. The length of the court shall be a hundred cubits, and the breadth fifty, and the height five cubits, with finely woven linen and bronze sockets throughout. All the equipment needed for serving the Tabernacle, all its pegs and those of the court, shall be of bronze (anon 1974).

Arsenal at Zeia, Greece (347-6 BC)

Specifications for several Attic (Greek) construction projects survive, inscribed on stone. Two projects in particular are well-covered by these inscriptions. Coincidentally, Vitruvius tells us something of them. They are the naval arsenal at the port of Piraeus, and the prostyle entrance hall to the cella of Ceres and Proserpine at Eleusis (several specifications for this project survive). Vitruvius states that they are both by the same architect, Philo, and that Philo published a book on the arsenal, and a book on the proportions of temples.

Again, these specifications are likely to have been very influential over time. They were 'published' in Greece, and Philo's books may well have contained or discussed them. Vitruvius and his contemporaries were familiar with the books and Greek specification practice, and he in turn influenced later Roman architectural writers, was known in the Middle Ages, and influential during the Renaissance and thence down to our own time. So we can trace modern specification practice back to at least these early Greek specifications.

The specification given here, for the arsenal, is an elaborate description (text on shelving and chests has been omitted) which lends support to the argument that drawings were not used contractually (other inscriptions record the use of paper – papyrus – for other projects), though models, measurements and architect's directions are mentioned. Indeed, the arsenal can be reconstructed from the description, a good understanding of Greek construction methods being a prerequisite (see Jeppesen 1958; Kostof 1977). The arsenal, to contain the movable equipment of the Athenian navy, was built between 340 and 330 BC. There was probably a separate contract. The building does not survive. The Attic foot (podon), used in the original, is 350 mm. It was divided into four palms (palastai), each divided into four digits (dactylai). A plethron was 100 Attic feet (Dilke 1987).

Gods.

Specifications for a stone tackle-store for hanging tackle, by Euthydomos Deimeitrios from Melite and Philonos Exeikestidos from Eleusinis.

Build a tackle-store for hanging tackle at Zeia, starting after the propylaion [vestibule or entrance to a temple area or other enclosure] from the agora coming from behind the dry-docks with one roof, length of four plethra width of fifty podai and five including the walls.

Excavate the space, depth from the highest point of three podai, clear the rest. Upon solid ground lay the foundations, and trim level, the tops all straight by the water-level.

And lay foundations then for the columns, clear from either wall by five and ten podai including the thickness of the columns, number of columns in either row of five and thirty, leaving a passage for the public through the middle of the tackle-store, width between columns of twenty podai. Make the thickness then of the foundations of four podai. Lay the stones alternately cross-wise and along the length.

Then build the walls of the tackle-store and the columns from Aktitis stone. Lay the euthynteirion [roughly, the footings or plinth] for the walls, width of three podai thickness of three half-podai length of four podai, then stones upon the corners, of four podai and three palastai.

And upon the euthynteirion lay uprights around the middle of the euthynteirion, length of four podai thickness of five half-podai and a daktylon height of three podai. Upon the corners the length is from the measure of the triglyphs.

Leave doorways in the gable-ends of the tackle-store, two at each end, width of nine podai, and build a partition at each end between the doors, width of two podai then inwards ten podai. And return the walls as far as the first columns so as to make an opening-stop for either door.

Then upon the uprights build walls of blocks, length of four podai width of five half-podai, but upon the corners the length is from the measure of the triglyphs, thickness of three half-podai. Then make the height of the walls from the euthynteirion of seven and twenty podai, including the triglyphs under the cornice.

The doorway height is five and ten podai and a half-podai. And lay upon them lintels of stone from Penteleikon, length of twelve podai thickness equal to the wall, height two courses. Erect pilasters of stone from Penteleikon or Hymeittion, lay under thresholds of Hymeittion. And lay a cornice upon the lintels, projecting three half-podai.

And make openings all around in all the walls opposite every inter-columniation, then in the gable-ends three at each side, height of three podai width of two podai. And enjoin to each opening a bronze grille, fitting. And lay upon the walls a cornice all around. And build pediments and lay a cornice down upon the pediments.

And erect the columns, lay under them stylobates, tops level with the euthynteirion, thickness of three half-podai width then three podai and a palaste length of four podai. Thickness of the lower part of the columns of two podai and three palastai, length including the capitals of thirty podai, every one of seven drums, length of four podai but the first five podai. Lay capitals upon the columns of stone from Penteleikon.

And lay main beams of wood upon the columns, fastened together, width of five half-podai height of nine palastai at the highest point, total eighteen in either row. And lay cross beams upon the columns above the passage, width and height equal to the main beams. And lay upon them principal rafters, width of seven palastai height of five palastai and two daktyloi, set to the roof-slope. Support them on struts upon the cross-beams, length of three podai width of three half-podai. And rebate for the principal rafters where supported upon the cross beams. And lay upon them purlins, thickness of ten daktyloi width of three palastai and three daktyloi, leaving a gap from each other of five palastai. And lay upon them common rafters, width of a half-podon thickness two daktyloi, leaving a gap from each other of four daktyloi.

And lay upon them boards, thickness of a daktylon width of six daktyloi, cover, fasten with nails of iron. Tile with Korinthian tiles fitted against each other.

And lay above the doorways upon the front parts on the inside a stone soffit with stone from Hymeittia. And upon the doorways of the tackle-store fasten doors. Make them of bronze outside. And cover all the floor with stones inside, joined together against each other, and work it straight and level on top.

...

So that there may be fresh air in the tackle-store, when building the walls of the tackle-store leave gaps in the masonry at the joints or as directed by the architect.

In this way everywhere shall work be carried out by the contractors conforming to the specification and from the measurements and from the model as directed by the architect. And they shall contract to timely delivery for every work item.

(Translation from Greek by the author based on the transliteration provided in Jeppesen 1958.)

Civil works, the Fayum, Egypt (245 BC)

The original of this specification is in Greek – Egypt was under Ptolomaic rule (323-30 BC) at this time – and on papyrus. The works described are in the Fayum district. The first sentence assigns a date to the document. The revetments were hurdles of reeds, intended to consolidate the mud banks to support wooden bridges. The widths were probably those of replacement bridges. The language is standardised and repetitive, as was the work. There were no drawings. This sort of work was done annually and was well understood. Contractual and descriptive material is mixed in the one document. The state client provides the tools, not an uncommon provision. Perhaps because the penalties are so severe, the work was undertaken by a government employee rather than by a private contractor. A peicheon was the royal cubit, of 523 mm, divided into 28 digits. A schoinion was 100 cubits (Hunt and Edgar 1977).

In the reign of Ptolemy, (son) of Ptolemy and Arsinoe the gods Adelphi, 2 [i.e "year #2"]. The priest to Alexander and the gods Adelphi (being) Tleipolemou (son) of Artaptes, the kaneiphorus of Arsinoe Philadelphus (being) Ptolomais (daughter) of Thyion in the month … at Krokodel-polis in Arsinoite is given out from the royal-treasury under auction through Ermaphilus the oikonomus in the presence of Theodorus the architect (and) Lon… by Petosiris the royal scribe.

Take down two bridges at Ker … and revet against the under-parts and make-through the top to width p 8 [i.e. 8 peicheis]. And revet along … from the bend, for schoina 35. Take down two bridges at … and replace revets against the under-parts and make-through the top to width p 14. Take down the bridge at Iera Neisus and replace revets against the under-parts and make-through the top to width p 8. And revet where eaten away under water, for schoina 5. At Iera Neisus take down revets and replace revets at the location of the weakest parts, for schoina 15. Take down the bridge at Thmoinete and replace revets and make-through the top to width p 14. Take down the bridges at Phnebge and leading towards Chanaanais and replace revets and make-through the top to width p 8. Work the bridge at Boukoton the same way. Take down the bridge at Phylakitike Neisus and replace revets against the under-parts. Work the bridge upon the water-way for Pterophorion farm the same way; work the three corners upon the canal leading towards Lysimachis the same way. Work the bridge against Kalliphanes farm the same way. Work the bridge upon the water-way through … for Tebetnois the same way. Work the bridge for Kamines the same way. Work the bridge at Kynon-polis the same way. And drive down strong piles to a length below the foundations p 1, apart from each other p (1). And entwining with rushes revet all the works with brushwood … height p 2. And make-through the top of all outlets width p 8.

Then guarantees of sufficient condition (being given) to the oikonomus, (you will) receive half the contract-sum, then when (you) work the value of the given money, (you will) receive the rest … (you will be) supplied then from the royal-treasury with as many mattocks as necessary, (you will) return these at completion of the work weighed at the same farm. And … transport yourself. Then if you do not work or conform to the written terms, it is permitted for these (officials) to properly auction again the work and contract by day labour. And whatever the full (costs) are found (to be) for auctioning again or contracting by day labour, (you will) return forthwith these expenses, the contract-sum, the money already paid (increased) by half, and damages, and about yourself the king (will) decide.

Undertaken by Oros the nomarch.

(Translation from Greek by the author based on the transliteration provided in Hunt and Edgar 1977.)

Vitruvian specifications, Rome, Italy (24 BC)

No Roman specifications survive (MacDonald 1982). Perhaps the closest we have are 'pseudo-specifications' in Vitruvius. Several are reproduced here. These read like extracts from project specifications, addressing the contractor rather than the architects who were the main audience for his book. They perhaps give us some idea of what a Roman specification would cover, and how it might read. One might infer that the Roman architect was also the contractor, but this is not so. The two were quite separate at the time. Also included is Vitruvius' description of his basilica at Fano, which Granger (1998) argues is "obviously taken from something like a specification". It is similar in content to the example of the arsenal at Zeia, above. Vitruvius tells us that contracts and drawings were prepared, to supplement such descriptions. Dating from the late second century BC, the Roman foot (pes) was 296 mm, divided into 12 inches (unciae) (Dilke 1987).

After slaking (the lime), mix your mortar, if using pit sand, in the proportions of three parts of sand to one of lime; if using river or sea-sand, mix two parts of sand with one of lime ...

In the middle, the main roof between the columns is 120 feet long and sixty feet wide. Its aisle round the space beneath the main roof and between the walls and the columns is twenty feet broad. The columns, of unbroken height, measuring with their capitals fifty feet, and being each five feet thick, have behind them pilasters, twenty feet high, two and one half feet broad, and one and one half feet thick, which support the beams on which is carried the upper flooring of the aisles. Above them are other pilasters, eighteen feet high, two feet broad, and a foot thick, which carry the beams supporting the principal raftering and the roof of the aisles, which is brought down lower than the main roof ...

Then, upon this (fern-covered planking) lay the bedding, composed of stones not smaller than can fill the hand. After the bedding is laid, mix the broken stone in the proportions, if it is new, of three parts to one of lime; if it is old material used again, five parts may answer to two in the mixture. Next, lay the mixture of broken stone, bring on your gangs, and beat it again and again with wooden beetles into a solid mass, and let it be not less than three quarters of a foot in thickness when the beating is finished ...

Having finished the mouldings, apply a very rough rendering coat to the walls, and afterwards, when the rendering coat gets pretty dry, spread upon it the layers of sand mortar, exactly adjusted in length to rule and line, in height to the plummet, and at the angles to the square (Morgan 1960).

Cathedral at Gaza, Palestine (402-7)

This Greek document, from Mark the Deacon's *Life of Porphyry* (a bishop of Gaza), makes it clear there had been major changes by this time, even within the Roman empire. The Roman construction industry had collapsed. Ecclesiastical buildings were often constructed in an ad hoc manner, with no room for specifications and contracts but with at least a sketched drawing. Site preparation of this building had begun before the plan had even been finalised.

Acquisition of materials and labour was ad hoc, too. It is likely that another building based on the same plan would have turned out entirely differently – there was no certainty in this approach.

As for the holy Bishop, he said, "Let us leave this [the plan of a church to be built on the site of the burnt Temple of Zeus Marnas], too, to the will of God." And while the site was being cleared, there arrived a magistrianos with imperial letters of [Empress] Eudoxia of eternal memory. ... On another sheet enclosed in the letter was the plan of the holy church in the form of a cross, such as it appears today by the help of God; and the letter contained instructions that the holy church be built according to this plan. ... Furthermore, the letter announced the despatch of costly columns and marbles.

... the holy Bishop ordered that the remaining debris from the marble revetment of the Marneion ... would be used for paving the open space in front of the church ...

The holy Bishop had engaged the architect Rufinus from Antioch, a dependable and expert man, and it was he who completed the entire construction. He took some chalk and marked the outline of the holy church according to the form of the plan that had been sent by the most-pious Eudoxia. And as for the holy Bishop, he ... commanded the people to dig. ... And so in a few days all the places of the foundations were dug out and cleared. ...

The following year the empress Eudoxia despatched the columns that she had promised – thirty-two big and admirable columns of Carystos marble which now shine in the holy church like emeralds. As these arrived by sea ... everyone ... brought carts and, after loading the columns one by one, pulled them along and deposited them in the open space of the church (Mango 1972).

Monastery at Cluny, France (1040-50)

The Farfa Consuetudinary is perhaps the nearest thing we have to a specification from this time and place. In spite of its name, it is thought to describe Odilo's monastery at Cluny, Cluny II, dated c. 1042-1049. Some parts of the document describe work completed, others (such as the extract given here) describe work to be done. There seem to have been no drawings. The intention seems to have been to provide enough information, in a letter, to the people at Farfa for them to develop a similar scheme themselves. The Carolingian foot was, at St Gall (816), 340 mm. Other sizes were also used e.g. 333 mm (Horn and Born 1979).

Next to the galilee is due to be built a palatium of length 130 and 5 feet, for the receiving of all unexpectedly-arriving men, who arrive with riders at the monastery. On one part of this house will be prepared 40 beds and just as many cushions and covers where rest men only, with latrines 40. On another part will be arranged 30 little beds where countesses or another honest women rest, with latrines 30, whereto alone they look after their needs. In the middle however of this palatium will be affixed tables, just like the boards of the refectory, where eat both men and women. In great festivities this house will be adorned with hangings and covers and with coverings on the seats themselves.

In front of this will be another house of length ft. 45, of width 30. For this in length extends up towards the sacristy, and there will sit all the tailors and also the shoemakers for making that which the chamberlain instructs; and for preparation they will have there a table of length 30 ft., and another table will be affixed with it, so that in width together the tables will be 7 feet. For between that mansion and the sacristy, and also the church and also the galilee will be a cemetery, where the laiety will be buried.

From the southern gate all the way towards the northern gate, on the west, will be constructed a house of length 280 ft., of width 25; and there will be constructed stables of horses divided into cubicles, and above will be a solarium where servants will eat and also will sleep, and they will have tables arranged there of length 80 ft., of width 4. And however many arrivals are not able to be refreshed at the mansion which we described above, they may be refreshed at this place.

(Translation by the author using the transliteration provided in Braunfels 1972.)

Church at Cordelier, France (1284)

This 'estimate' (devis) for a church at Cordelier in Provins, France would probably have complemented drawings, according to Gimpel (Waugh 1983). The 'estimate' gives the dimensions, such as they are, rather than the plans. An initial sentence seeking blessing is still common in the contracts of some Moslem countries, such as Iran. The description relies heavily on custom, for determining the size of the arches for example. Elevations were therefore not needed, and in this case the remains of the convent provide a plan of sorts.

In the name of the Father, the Son and the Holy Spirit, Amen. This is the estimate for the convent of the Frères Mineurs at Provins. Firstly the convent shall be razed to the ground. The front gable and the side will be the same size they were before but the side of the wing will be built on round pillars with arches the same length as the old wing. The height of the curve of these arches will be determined by their width to allow them to join the entablature which will support the structure of the nave; and there will be, on this side, as many arches as there will be bays between the tie-beams, all dictated by the length of the old wing; and the size of the arches will be governed by the structure. On the courtyard side, where the springing line meets the gable, there will be a six-foot-deep buttress with a three-foot bond-stone, which will be cushioned by a chamfered facing above the support of the gable (Waugh 1983).

King's College Chapel, Cambridge, England (1440)

Following is an extract from Henry VI's 'will' (a statement of the royal will or desire or command) describing King's College Chapel at Cambridge – essentially a brief for a design and construct contractor. The 'will' goes on to describe the quadrangle, library, hall, lodge and walls. Henry founded the College in 1440-41, to support a rector and twelve scholars. By 1445 however, he had expanded his vision to a college of seventy scholars, under a provost. The chapel was the only portion of Henry's design that was completed.

Henry wished the college as a whole to be built "in large fourme, clene and substancial, settyng a parte superfluite of too gret curious werkes of entaille and besy moldyng". This is a statement about the particular sort of Gothic (Perpendicular) that Henry wanted – a bit like specifying a deconstructivist design, in contemporary terms! There was no specification in the modern sense.

The master masons (principally Reginald Ely) and master carpenters were actually responsible for the design and execution of the chapel under the financial direction of a succession of masters of the works, and indeed the 'will' was varied in a number of ways. For example, the octagonal corner turrets of the chapel are not mentioned in the 'will', which also anticipated only eight side chapels ("closettes") whereas 18 were built. The vestry wasn't built at all. Clearly the builders were given a deal of latitude in the interpretation of the brief.

The building was detailed as work progressed, with the result that vaulting changed mid-stream from lierne to fan, leaving some redundant mouldings in the vaulting-shafts. The timber roof wasn't drawn until 1480, when a pennyworth of parchment was bought for the purpose. A series of specified contract packages, with drawings, followed e.g. for the stone roof, for the finials of the buttresses, for the vaulting of porches and chapels, and for glazing. Progress was intermittent at best, a common feature of royal works of the time. The medieval English foot was 305 mm (Colvin 1963).

And as touching the dimensions of the church of my said College of our lady and saint Nicholas of Cambridge, I have devised and appointed that the same church shall contain in length 200+4x20 8 [CCiiij^{xx} *viij i.e. 288] feet of size without any aisles and all of the width of 40 feet and the length of the same church from the West end unto the Alters at the choir door, shall contain 120 feet. And from the Provost's stall unto the steps called "steps of the choir" 4x20+10 [i.e. 90, equivalent to the modern French, quatre vingt dix] feet for 36 stalls on either side of the same choir, answering unto 70 fellows and 10 priests hired which must be of the first type; and from the said stalls unto the East end of the said church 62 feet of size. Also a reredos bearing the Rood-loft separating the choir and the body of the church, containing in length 40 feet, and in breadth 14 feet; the walls of the same church to be in height 4x20 10 [i.e. 90] feet, embattled vaulted and waggon roofed sufficiently buttressed, and every buttress finished with finials.*

And in the East end of the said church shall be a window of 11 lights, and in the west end of the same church a window of 9 lights and between every buttress a window of 5 lights. And between every of the same buttresses in the body of the church, on both sides of the same church, a closet [i.e. side chapel] with an alter therein, containing in length 20 feet, and in breadth 10 feet, vaulted and finished under the sill of the aisle windows: and the pavement of the church to be enhanced 4 feet above the ground without, and the height of the pavement of the choir 1 foot half above the pavement of the church, and the pavement at the high alter 3 feet above that.

Item, on the north side of the choir a vestry containing in length 50 feet, and in breadth 22 feet, separated into 2 houses beneath and 2 houses above, which shall contain in height 22 feet in all with an entry from the choir vaulted.

Item, at the west end of the church a cloister square, the East pane containing in length 175 feet, and the west pane as much; and the north pane 200 feet, and the south pane as much, of the which the deambulatory 13 feet wide, and in height 20 feet to the corbel-table, with clerestories and buttressed with finials, vaulted and embattled, and the ground thereof 4 feet lower than the church ground; and in the middle of the west pane of the cloister a strong tower square, containing 24 feet within the walls, and in height 120 feet unto the corbel-table, and 4 small turrets over that, finished with pinnacles, and a door into the said cloister ward, and outward none.

(Modernised by the author based on the transliteration provided in Willis and Clark 1988.)

Artisans' houses, Amsterdam, The Netherlands (1670)

This extract describes houses to be built in the Noortsche Bosch, Amsterdam. Like all the specifications quoted here, this does not necessarily represent good or best contemporary practice. It may not even be typical. Contractual material is included. Clauses 29-34 are missing in the source, but describe penalties and so on and the undertakings of the contractors. Text on piling and roofing has been omitted here.

The requirement that builders provide all their own tools is the modern norm, but as we have seen was not always standard historically. The provision of beer (the water was not safe to drink) was common enough, from as far back as the Egyptian period. A corresponding clause in the related brickwork specification makes it clear that the beer is for the contractor's workmen. Provision of some items by the principal remains widespread. Note the reference to two classes of drawings, those supplied by the principal, and those to be prepared by the contractor.

Dutch feet (voet), of 280 mm, were divided into 11 inches (duijms) (Zantkuijl date not known).

Timber work

...

5. The contractor shall help to fetch all this sawn wood to their work from the yard, where the millers have it on board.

6. All the front and rear gable (and) also this small-kitchen are to be made with such frames, including all doors (and) windows, according to the drawings, of which the contractor shall hand-execute (a) paper copy; and also a cross-frame in the room overlooking the premises is to be made known [i.e. it has not been shown in the drawings].

7. In the cellar shall be provided door-frames, first in the wall, there separating the front room from the cellar, then second at the peat-store of the cellar, and the third at the stair from provision-cellar of the house. Which three doors in the drawings do not stand as directed.

[meaning obscure] And (to the) fronts as many doors, windows and frames, all to be done (as) directed in the drawings, or be done with.

8. In the living-cellar shall be built a bedstead, mantelpiece, first backed with two inch greenwood, and then clad with a spring-lining, frieze and architrave; same in the room above and in the kitchen also, a store cupboard in the kitchen. (To the) fronts (a) window (frame) and windows and doors in the whole house or work. Every one to be hung with such locking, all shall be delivered (by) the principals thereto, and that all according to requirements. All the windows together with inner – and outer – doors are to be made with glass panes.

9. In the room shall be built a bedstead with a spring-lining around, with attics above in the bedstead, closed grooved and dressed, of which also the bed-boards must be grooved in each others [sic], with a trundle-bed thereunder. The other bedstead in the kitchen. Also (a) wainscot, dressed with the bed-boards adjoining, and front what more thereon is missing [i.e. provide a front to the extent there is not one]. A store cupboard must be built, at where [i.e. in the location where] the principals shall indicate.

10. In the front room must be built such fashion of stair, all done as in the presentation drawings, and also (there) is to be made a winding-stair around from the front room into the provision-cellar; also a few boards there must be for a beer-rack for convenience of said cellar. Also a steep-stair, going from the room to serve the attic. Which stairs must every one be built to the fashion and according to requirements of the work. Under said stair comes a door; frame this with an architrave mounting, all must at this frame from the kitchen and to the room adjoining [meaning obscure].

11. In each floor up to three shall be placed a layer of beams; the two first floor(s) from 8 and 12, the top-most layer from 11 and 7 inch, with their treveel-timbers and stair-gates, worked all according to requirements. All the ends of the beams must be nailed by the contractor, for which the nails shall be delivered to him by the principals.

12. The first floor with planks (which) are to be dressed, front coated, grooved and feathered, and well fast-nailed [fastened] according to requirements. Second and third square adjoining (faces) dressed and fronts (made) all as the others. The outside on which window-fasteners must be nailed. This expel, that shall come from the principals until installed [meaning obscure]. All like this the second attic from the kitchen is also to be done.

13. The kitchen has two layer(s) (of) beams; shall be placed on the flat from 7 and 11 inch. The lowest floor with planks, also must the above-attic (have); all from best condition all (as) the other. Said kitchen must have a steep-stair going into the yard, that the contractor also must make, and in the kitchen is to be made, all (by) him, (the) contractor, shall be made as directed, a peat-case [i.e. a store for peat, or fuel].

…

17. This roof of the great-house and of the kitchen is to be covered with 2 1/2 and 3 inches ribs, defect-free and nailed according to the requirements, all then clad with beam-laths, placed by the bricklayers and born by the timber-layers and to be nailed according to requirements. This whole roof overall and what thereon is fixed, must be dressed.

18. The contractors must surround all the premises, in every particular, according to fashion with a shooting fence and its pieces in each other grooved with deep-grooves. The pales on every side are to be brought, all that (are) required.

19. In each house must be built a privy-house on a common pit, with glazing and to requirements and a door hung. All well dressed and built according to requirements.

...

21. Also the bottom of the privies and of the rainwater-tanks must be built, according to requirements of (the) agreement.

22. This beer shall be delivered (by) the principals; that is eight cask(s) for each house in the estate.

23. All the materials of this aforesaid work shall be delivered (by) the principals themselves.

24. All these tools that they need for building of the aforesaid houses, not poor quality, the contractors shall provide themselves.

25. All the outer-frames are to be made with a groove and a keel-wood there-in [therein] from one inch lath; the frames are to be set two inch(es) into the wall.

26. Also in the rear gable is to be made a cross-frame and three sill-frames.

27. The attic-panels the contractor shall make to both faces well shaved, grooved, inlaid and properly made fast.

28. In the kitchen is to be made a shopfront with three lights a door-frame, and under in the cellar two lights and a sill-frame with a window.

(Translation by the author based on the transliteration provided in Zantkuijl, with help from John Schooling and colleagues at DEM, Sydney, and Rob Visser at OECD, Paris.)

House, North Carolina, USA (1788)

The planter William Lenoir replaced his log dwelling with one of the first well-finished large houses in the district, and wrote a detailed contract with the carpenter Thomas Fields. The house was timber framed, lined and clad, so this contract pretty well covered the job. The typical contract at that time and place simply specified that the work be finished in "a good workmanlike manner what belongs to a carpenter & joiner" (Bishir et al 1990). What follows, then, though brief, is for a house of first-quality craftsmanship. Note that the quantities are specified. Windows would be sized using standard glass panes, 200 x 250 mm. The associated plans for his two-storey 12.2 x 7.3 m house were rudimentary single-line sketches with room sizes shown as "24 x 18" (feet), for example. The carpenter was to

... frame said house with proper Girders and every thing that (is) necessary thereto – To lay all the lower Floors in the first & second stories with Plank Tongued & Gruved – To do all the weather–Boarding Plain'd Jointed & Beeded &c – To sheat the Roof with square edged Plank, and to get good thick shingles from the Stump, Joint them to a Guage, round them and nail

them on, cornish said house with Mondilions &c – make sixteen 18 light window Frames worked out of the solid wood & folding shutters of six pannels to each window the sashes &c. – make Ten 8 light windows & shutters of Two pannels Each & sashes &c. – make nine 6 pannel Doors & Frames &c – run up one pair of Stairs well Bannistered & a small pair cased up &c. – put Wash Boards and chair-Boards in the nicest manner Round all the Rooms and hang all the Doors and windows – Which is all to be Done in a proper good nice and work-man-like-manner (Bishir et al 1990).

Fort Denison, Sydney Harbour, Australia (1855)

This is an extract from the specification by the architect, William Cole. Note the reference to a drawing, yet the extensive use of dimensions. The drawings used must have been a curious mix of outlines (for the mason) and details (for the smith). Note also the two references to "established patterns" – presumably military standard details of some sort. These predate civil standards of the sort we use today. Indeed military standards, for guns and ammunition, have existed since at least the late 18th century, when Jean Gribeauval standardised French artillery production (Keegan 1993).

Mason – Front and end walls to be of ashlar masonry, 12 ft thick at bottom, and 9 ft at top, the headers not to be less than 4 ft 6 inches long, nor the stretchers less than 2 ft 6 inches on bed, the core to be filled in with rubble masonry, grouted with hot lime, the outer face to be pitched on rough rock work, the insides of the parapet to be picked work, the upper slope and embrasure to be wrought in chiselled work, the back wall to be of ashlar masonry, 5 ft thick at bottom and 4 ft at top, with pitched face to correspond with wall of barrack, the walls to be properly bonded into the old work; this wall to be carried up to form a parapet 4 ft 6 inches high above the terreplein, the top to be sloped outwards and finished in chiselled work; banquette of ashlar masonry, 1 ft 6 inches wide and 1 ft 4 inches high, to be fixed round parapet to correspond with that at present fixed; gun pivots to be sunk into stone blocks 4 ft x 4 ft, and run with lead; gun races to be let into flagging or curbs, flush and run with lead; the circle at south end to be paved with 8 inch flagging laid in mortar; surface drain to be of ashlar masonry, 2 ft 6 inches wide and 9 inches deep, with radiated joints, properly sunk and set in mortar.

Carpenter – Traversing platforms to be constructed, according to established pattern.

Smith – Gun pivots, of cast iron, to be made according to established pattern, and gun races of 4 inch x 3/4 wrought iron, as per plan (Kerr 1986).

House, Lane Cove River, Australia (1891)

This is an extract from an extensive manuscript specification by William Martin, architect, for a weatherboard house at Woolwich, Lane Cove River, NSW, dated 1891. The client was H Carpenter of the Department of Public Works. Elsewhere this specification includes contractual material even though there is a separate contract. Drawings are referred to. Note the reference to Martins cement (perhaps made by the architect's family), and the cross-reference to another page in the specification. The unhelpful term "unless otherwise specified" appears twice in this short extract. Note also the various approvals and directions – all hard to price at tender. Use of prime cost sums and 'to approval' (really forms of procrastination rather than a method of specification) are still widespread.

PLASTERER

Pointing

Point all brickwork and flashings, and weather in cement chimney tops and projecting upper course.

Ceilings

All ceilings unless otherwise specified to be lathed with split laths and plastered three coats best hair mortar rendered and set trowelled to a good face left white and perfect and well keyed.

Internal walls

Lath with split laths plaster float and set all interior walls unless otherwise specified, the setting coat to be coloured as directed. Render float and set all interior brickwork.

Arch in hall

Run arch and jambs of same in hall moulded in Martins or other approved cement.

Cornices

Run moulded cornice, design to be furnished, to two front rooms and intervening front hall 2'6" in girth. Vide page 29.

Centre flowers

Enriched centre flowers 3 ft in diameter to two front rooms, with strong iron hook concealed in centre firmly screwed to ceiling joists, for swinging lamps, hall to have enriched centre 18" diameter with hook in centre same as front rooms. Centre flowers to be approved by Proprietor.

Construction of plaster

All stuff for ceilings and walls is to be best shell lime and washed clear sharp sand in equal proportions with good dry hair mixed in pricking up coat.

Ventilations

Two ventilators to be placed where directed in each room except bath and pantry which are to have one each, size of openings 7" x 10" to have light cast iron grating on outside face and approved ornamental cement or plaster grating on inside, the outside to be protected from weather by light zinc hood of approved pattern. In the rooms lined with wood the interior of ventilators to be composed of perforated zinc surrounded with a neat pine bead to cover joint. Pierce ventilators as shown on front elevation in each gable end to be properly lined with wood and covered half way through wall with perforated zinc, the opening to be neatly beaded round and sills well sloping and weathered.

Mantlepieces and grates

Provide and fix of value below mentioned free of all commissions in Sydney grates and mantlepieces to be selected by Proprietor where he chooses and carriage paid by contractor – all inclusive of fixing.

Tile hearths and margins, marble fender, mantlepieces, grates

Two front rooms to have tile hearths and back hearths and tile margins valued 10/- per foot square exclusive of fixing, and marble fender rail and plinth to mantlepiece value £2.10.0 each prime cost and grates valued at £1.10.0 each and mantle pieces exclusive of marble plinth to cost £3 each. Bedroom grate 25/- and mantlepiece £2.10.0.

Range

Provide and fix in kitchen a cooking range valued at £6.10.0 exclusive of commission carriage and fixing to be selected by the Proprietor where he chooses with all necessary flues etc (source unknown).

House, NSW, Australia (1992)

This extract from an anonymous NSW specification, dated June 1992, is not based on NATSPEC material. Note the reference to Fortecon and Bulli bricks – proprietary specification is commonplace now. The specification also refers to standards, in particular, to "relevant" standards and to unidentified standards. Drawings are also referenced in a very general way. These sorts of cross-references are not helpful. A prime cost sum is used and again, this is unhelpful. The style is more verbose than the sample specification in Appendix 4. The engineer's specifications mentioned would typically be on the drawings (the first sheet in the structural set), and may not be project-specific.

CONCRETE AND STRUCTURAL STEEL

All concrete work to be in accordance with AS3600 "Concrete Structures" and Engineer's Drawings and specifications. All structural steel work to be in accordance with AS4100 "Steel Structures" and Engineer's Drawings and Specifications.

Underlay

Under all internal slabs on fill place a layer of "Fortecon", lapped 150mm at edges and taped. Securely tape all around the perforations.

Concrete finishes

Where concrete is specified topped, paved or tiled, leave rough for key. Otherwise grade to outlets and finish with a broom.

Carport and Driveway slabs to have selected colour additive sprinkled on top of newly poured concrete and mixed in to give an even finish before concrete sets.

Carport, garages and all external concrete to be graded for run-off. Footings: All new brickwork to be built on 400 mm wide x 75 mm deep concrete levelling strip on rock.

Concrete Slab under New Spa: Concrete slab to be minimum 100 mm thick with F62 mesh in top. Cover over mesh: 25 mm concrete. Concrete strength 20 MPA at 20 days.

Extent of Work

As shown on Drawings.

BRICKWORK AND CEMENT RENDER

Read in conjunction with Engineer's Specification.

Make good after demolition where necessary.

Generally

Materials and workmanship to conform with the relevant A.S. Codes.

Brickwork shall be laid in stretcher bond, reinforced every fourth course with 75mm galvanised welded brick reinforced mesh, all in accordance with Australian Standard.

Build in all damp courses, bolts, lugs, ties, hoop iron straps, metalwork and plates as work proceeds.

Wall ties shall be 10 gauge galvanised at 900mm centres every fourth course and staggered.

All 230mm walls to be built fully bonded in English Bond.

Bricks

Face bricks

To match existing to all external skins and to internal walls shown as facework.

All new work to be brick-veneer construction to match existing.

A PC Sum is allowed for the supply only of selected face bricks for all visible internal and external work.

Builder to allow for mortar and all other incidentals.

Bricks shall be Bulli 'Light Sandstock' laid as facework for all visible external and internal work.

Appendix 2 — AACA performance criteria

This Appendix supplements the discussion on vocation in Chapter 2.

The nine performance criteria for the Architects Accreditation Council of Australia (see AACA 1993) Competency Standards Elements 2.2.3 and 2.2.4 (under Unit 2, Context 2.2) are as follows:

Unit 2 Documentation

Context 2.2 To generate documentation of a building project so that it can be costed, built and completed in accordance with the brief and time, cost and quality objectives.

Element 2.2.3 Prepare architectural specifications and schedules.

Performance Criteria:

- specifications and schedules are produced which are timely, correct, complete and comprehensible to architects, consultants and contractors
- specifications nominate quality and performance standards with regard to selected materials, finishes, fittings, components, systems and special items
- specifications declare the type and extent of work of separate building trades and subcontractors
- specifications, schedules and drawings are cross-referenced and co-ordinated
- specifications and schedules from consultants are verified for matching with architectural documentation
- schedules nominate the type, quality and performance standards of building components and special items
- specifications and schedules are consistent with the design intent, in terms of quality and performance standards and conform with the codes and standards of regulating bodies

Element 2.2.4 Co-ordinate the documentation of the project.
Performance Criteria:

- architectural and consultants drawings, specifications and schedules are reviewed and confirmed as consistent and compatible with each other, with the design intent and with quality, cost and time parameters
- all drawings, specifications and schedules are consistent with the type of building contract and/or procurement procedure which has been selected for the project

Appendix 3 — Conditions of contract — Extracts

This Appendix, comprising selected extracts from some of the general conditions of contract in use in Australia, supplements discussion on coordination with conditions of contract, in Chapter 5. The extracts are those most relevant to specifiers and provide a useful cross-reference between similar clauses in different contracts. See Table 5.2. Omissions from each extract are indicated thus: ***

Extracts from AS 4000 General conditions of contract (1997)

1 Interpretation and construction of Contract

***

legislative requirement includes:

a) Acts, Ordinances, regulations, by-laws, orders, awards and proclamations of the jurisdiction where *WUC* [i.e. work under the contract] or the particular part thereof is being carried out;

b) certificates, licences, consents, permits, approvals and requirements of organisations having jurisdiction in connection with the carrying out of *WUC*; and

c) fees and charges payable in connection with the foregoing;

***

8 Contract documents

8.1 Discrepancies

Figured shall prevail over scaled dimensions in a discrepancy. Otherwise, if either party discovers any inconsistency *** that party shall give the *Superintendent* written notice of it.

***

11 Legislative requirements

11.1 Compliance

The *Contractor* shall satisfy all *legislative requirements* except those in Item 19(a) [a prompt in the Annexure] or directed by the *Superintendent* to be satisfied by or on behalf of the *Principal*.

***

20 **Superintendent**

***

Except where the *Contract* otherwise provides, the *Superintendent* may give a direction orally but shall as soon as practicable confirm it in writing.

***

28 **Materials, labour and construction plant**

Except where the *Contract* otherwise provides, the *Contractor* shall supply everything necessary for the proper performance of the *Contractor's* obligations and discharge of the *Contractor's* liabilities.

***

29 **Quality**

29.1 **Quality of material and work**

Unless otherwise provided the *Contractor* shall use suitable new materials and proper and tradesmanlike workmanship.

***

30 **Examination and testing**

30.7 **Costs**

Costs in connection with testing pursuant to this clause shall be borne by the *Principal* except where the *Contract* otherwise provides or the test is consequent upon, or reveals a failure of the *Contractor* to comply with the *Contract* (including this clause).

***

Extracts from AS 4300 General conditions of contract for design and construct (1995)

***

4 **Warranties**

4.1 **Contractor's warranties**

*** the Contractor warrants to the Principal that the Contractor -

***

(d) shall execute and complete the Contractor's Design Obligations and produce the Design Documents ***; and

(e) shall execute and complete the work under the Contract in accordance with the Design Documents so that the Works, when completed, shall -

- (i) be fit for their stated purpose; and
- (ii) comply with all the requirements of the Contract and all Legislative Requirements.

***

8 Contract documents

8.1 Discrepancies

The several documents forming the Contract are to be taken as mutually explanatory of one another. *** In the event of an ambiguity or discrepancy being discovered *** the Superintendent shall direct the Contractor as to the interpretation to be followed by the Contractor in carrying out the work.

***

14 Legislative requirements

14.1 Complying with legislative requirements

The Contractor shall satisfy all legislative requirements except those which (a) are specified in Annexure Part A; or

(b) the Superintendent directs are to be satisfied by or on behalf of the Principal.

***

23 Superintendent

***

Except where the Contract otherwise provides, a direction may be given orally but the Superintendent shall as soon as practicable confirm it in writing.

***

29 Materials, labour and constructional plant

29.1 Provision of materials, labour and constructional plant

Except to the extent that the Contract otherwise provides, the Contractor shall supply everything necessary for the proper performance of the Contractor's obligations *** under the Contract.

***

30 Materials and work

30.1 Quality of material and work

The Contractor shall use the materials and standards of workmanship required by the Contract. In the absence of any requirements to the contrary, the Contractor shall use suitable new materials and proper and tradesmanlike workmanship.

***

31 Examination and testing

31.7 Costs of testing

***

Such costs shall be borne by the Principal *** unless -

(a) the Contract provides that the Contractor shall bear the costs, or the test is one which the Contractor was required to conduct *** ;

(b) the test shows that the material or work is not in accordance with the Contract; Where such costs are not to be borne by the Principal, they shall be borne by the Contractor ***

***

Extracts from SBW-2 lump sum contract for simple building works (RAIA/MBA 1997)

***

2.3 **Precedence**

2.3.1 Should any ambiguity, discrepancy or inconsistency exist in the Contract Documents then the order of precedence of Contract Documents set out in Item D of the Appendix shall apply to resolve the same.

***

5.3 **Architect's instructions**

Any instruction which may be given by the Architect shall be in writing, and signed by the Architect,[.] ***

Except where this Agreement otherwise provides, the Architect may give an instruction orally but shall promptly confirm it in writing.

***

6.1 **Construction methods**

The Builder is and shall remain responsible for all construction means, methods, techniques, sequences and procedures employed and to be employed by him in and about the execution of the Works and for co-ordinating all portions of and the execution of all portions of the Works and the Architect shall not be entitled to issue any instructions as to these matters.

***

6.3 **Requirements of authorities**

The Builder shall comply with all statutes, ordinances, regulations and by-laws of any authority having jurisdiction over the Works or with whose systems the same are or will be connected and shall *** pay all fees and charges legally recoverable in respect of the Works, and:

6.3.1 The Builder shall give all notices to and comply with all notices issued by any such authority.

6.3.2 Where prior to entering into this Agreement the Proprietor has given any notices, paid any fees or received any permits, approvals or other authorisations necessary to comply with the requirements of this clause then the Proprietor through the Architect shall provide to the Builder evidence of such.

***

6.6 **Materials and workmanship**

All materials and standards of workmanship shall be in conformity with this Agreement ***

***

6.6.3 Should the Architect instruct the Builder to *** arrange for or carry out any testing of any materials or goods *** or of any executed work then the cost *** shall be added to the Contract Sum unless:

***

.2 the opening up for inspection or testing of work or materials shows that such work materials or goods as the case may be are not in reasonable accordance with this Agreement; or

.3 any such cost has been included in the Contract Sum.

***

Extracts from Building works contract – JCC-C 1994 with quantities (1994a)

***

2.02 **Precedence**

Should the Contract Documents contain any discrepancy or inconsistency then:

2.02.01 The following order of precedence of Contract Documents shall apply to resolve the same:

.01 This document.

.02 The Specification(s).

.03 The Drawings.

.04 The Bills of Quantities, ***

***

5.04 **Written instructions**

Any instruction or notice which the Architect is obliged to give or chooses to give pursuant to this Agreement shall be given promptly and in writing.

***

6.01 **Scope of the works**

The scope of the Works shall include all the work specifically referred to in this Agreement, all Variations thereto instructed, sanctioned or required under the terms of this Agreement and all minor items of work obviously to be inferred therefrom for the proper execution and completion of the Works to the reasonable satisfaction of the Architect and for the proper performance by the Builder of his obligations under this Agreement.

6.02 **Provision of materials and labour**

Unless otherwise stipulated the Builder shall provide at his own expense everything necessary for the proper execution and completion of the Works and for the proper performance by the Builder of his obligations under this Agreement.

6.03 **Construction methods**

The Builder is and shall remain responsible for all construction means, methods, techniques, sequences and procedures employed and to be employed by him in and about the execution of the Works and for co-ordinating all portions of and the execution of all portions of the Works.

***

6.05 Requirements of authorities

The Builder shall comply with the requirements of the provisions of all Acts of Parliaments, ordinances, regulations or by-laws of all competent authorities and with the lawful requirements of public, municipal and other authorities which have jurisdiction over the Site or the Works and which arise by reason of the execution of the Works, including all Variations, subject to the following:

6.05.01 Where prior to entering into this Agreement the Proprietor has given any notices, paid any levies, charges or fees or received any permits, approvals or other authorisations necessary to comply with the requirements of this Clause 6.05 then the Proprietor through the Architect shall provide to the Builder evidence of such.

6.05.02 The Builder shall give all notices necessary to comply with such requirements and shall pay all levies, charges or fees in connection therewith which have not been given or paid by the Proprietor prior to entering into this Agreement.

***

6.08 Materials and workmanship

All materials used in the Works and standards of workmanship shall be in conformity with the provisions of this Agreement provided that:

6.08.01 In the absence of any such provision in this Agreement in respect of any material or standard of workmanship that material or that standard of workmanship, as the case may be, shall be of a kind which is both suitable for its purpose and consistent with the nature and character of that part of the Works for which it is used.

6.08.02 Any materials not otherwise specified shall be new.

***

6.08.05 Should the Architect issue instructions requiring the Builder to *** arrange for or carry out any testing of any materials or goods *** or of any executed work then the cost *** shall be added to the Contract Sum unless:

***

.02 the opening up or testing shows that such work, materials or goods are not in accordance with this Agreement; or

.03 any such cost has been included in the Contract Sum.

***

Extracts from Project contract PC-1 (PCA 1998)

***

6.2 Contractor's design

The Contractor must:

(a) design the parts of the Works which the Contract requires it to design and for this purpose prepare all relevant Design Documentation;

***

6.6 Fitness for purpose

The Contractor warrants:

(a) the Design Documentation it prepares will be fit for its intended purpose; and

(b) upon Completion the Works or each Stage will, to the extent they are designed by the Contractor, be fit for their intended purpose.

***

6.10 Resolution of ambiguities

If there is any ambiguity, discrepancy or inconsistency in the documents ***

(a) the order of precedence in the Contract Particulars will apply;

(b) where the ambiguity, discrepancy or inconsistency is between the Contract and any part of the Design Documentation, the higher standard, quality or quantum will prevail but if this does resolve the ambiguity *** the Contract will prevail;

***

8.2 All work included

The Contractor has allowed for the provision of all Plant, Equipment and Work, materials and other work necessary for the Contractor's Activities, whether or not expressly mentioned in the Works Description or any Design Documentation.

Any such Plant, Equipment and Work, materials and other work:

(a) must be undertaken and provided by the Contractor;

***

8.3 Statutory requirements

***

In carrying out the Contractor's Activities, the Contractor must:

(a) unless otherwise specified in the Contract Particulars, comply with all applicable Statutory Requirements;

(b) without limiting paragraph (a):

(i) apply for and obtain all Approvals *** ; and

(ii) give all notices and pay all fees ***

***

8.20 The environment

The Contractor must:

(a) ensure that in carrying out the Contractor's Activities:

***

(ii) it does not pollute, contaminate or otherwise damage the environment; and

***

(b) make good any pollution, contamination or damage to the environment arising out of, or in any way in connection with, the Contractor's Activities ***

***

9.1 **Construction**

The Contractor must in carrying out the Contractor's Activities:

(a) use workmanship:

(i) of:

(A) the standard prescribed in the Contract; or

(B) to the extent it is not so prescribed, a standard consistent with the best industry standards for work of a nature similar to the Works and which is at least to the same standard as that found in the development referred to in the Contract Particulars; and

(ii) which is fit for its purpose;

(b) use materials:

(i) which:

(A) comply with the requirements of the Contract;

(B) if not fully described in the Contract, are new and consistent with the best industry standards for work of a nature similar to the Works and which are at least to the same standard as those found in the development referred to in the Contract Particulars; and

(ii) of merchantable quality, which are fit for their purpose and consistent with the nature and character of the Works: and

(c) comply with the requirements of the Contract and, to the extent they are not inconsistent, the requirements of the Building Code of Australia and all relevant standards of Standards Australia.

***

9.4 **Testing**

The Contractor must carry out all tests required by the Contract or directed by the Contract Administrator.

***

9.5 **Costs of testing**

If:

(a) the Contract Administrator directs the Contractor to carry out a test which:

(i) is not otherwise required by the Contract; *** and

(b) the results of the test show the work is in accordance with the Contract,

the reasonable costs incurred by the Contractor in carrying out the test will be determined by the Contract Administrator and added to the Contract Price.

***

16.7 **Address for service**

Any notice to be given or served under or arising out of a provision of this Contract must be in writing **   nnn*

***

Extracts from C21 construction contract conditions (DPWS 1999a)

***

10. **Scope of the works**

.1 The scope of the Works is described in brief in Contract Information item 3 and more specifically in *Principal's Documents* and other *Contract Documents* (as applicable), and includes:

***

.2 all items not specifically referred to or described in the Contract which nonetheless are required to complete the Works and achieve the effective and efficient use and operation of the Works;

.3 all items of work reasonably inferred from the *Contract Documents* as necessary to properly execute and complete the Works; and

***

.2 The Contractor acknowledges that:

***

.2 it has made full allowance in the *Contract Price* for the matters referred to in clauses 10.1.1, 10.1.2 and 10.1.3 .

***

13 **Environment, safety and quality requirements**

.1 The Contractor is responsible for and must comply with the requirements of the Contract for:

.1 protection of the environment

***

14 **Statutory requirements**

The Contractor is responsible for compliance with all *Statutory Requirements*, subject to clause 42, except if (because of the nature of the requirement) only the Principal can comply.

***

16 **Licences and approvals**

The Contractor must obtain at its own cost all licences, approvals and consents necessary to carry out the Works in accordance with the Contract, and pay all fees and give all necessary notices arising out of Statutory Requirements, other than those the Principal has arranged.

***

19 **Notices and instructions in writing**

***

.2 All notices must be in writing, and all instructions by the Principal must be in writing or confirmed in writing as soon as practicable.

***

31 **Ambiguities**

.1 The Principal must resolve any ambiguities, inconsistencies or discrepancies in the Contract which are notified by the Contractor to the Principal for resolution.

***

32 Contractor's documents

.1 The Contractor must carry out and complete *Design* of the Works (whether or not it is responsible for the Principal's design under clause 33.1) and so produce *Contractor's Documents* which meet the requirements of all of the following:

- .1 the Contract
- .2 Principal's instructions
- .3 *Statutory Requirements*
- .4 the Building Code of Australia, and relevant Australian Standards
- .5 if no other standard is specified in the Contract, then good industry standards applicable to the Works.

***

.4 The Contractor must produce *Contractor's Documents* which will ensure that the Works and every part of them are fit for the purposes required by the Contract. This responsibility is reduced to the extent that the *Contractor's Documents* are not fit because of design work by the Principal for which the Principal retains responsibility.

***

37 Construction

.1 The Contractor must supply all *Materials* and construct the Works in accordance with all of the following:

- .1 the Contract
- .2 the *Contractor's Documents*
- .3 Principal's instructions concerning the Works
- .4 *Statutory Requirements*
- .5 the Building Code of Australia, and relevant Australian Standards
- .6 if no other standard is specified in the Contract, then good industry standards applicable to the Works.

.2 The Contractor must comply with this clause 37 so as to ensure that the Works and every part of them are fit for the purposes required by the Contract. This responsibility is reduced to the extent that the Works are not fit because of design work by the Principal for which the Principal retains responsibility.

***

38 Testing

.1 The Contractor must:

- .1 *Test* (at its own cost) all parts of the Works specified in the Contract to be *Tested*

***

Appendix 4 — Brick and block construction specification – Sample

This Appendix is from the *NATSPEC BASIC Templates*, January 2001.

BRICK AND BLOCK CONSTRUCTION

1 GENERAL

1.1 CROSS REFERENCES

General

Refer to the *General requirements* worksection.

Related worksections

Refer to the following worksections: >

1.2 STANDARD

General

Materials, construction and detailing: To AS 3700.
Masonry units: To AS/NZS 4455.

1.3 INTERPRETATION

Definitions

Face units: Masonry units used in facework, including purpose-made units such as squints, sills and thresholds.
Facework: Masonry in which the form, or form and colour, of the face units and joints is visible in the completed works.

2 QUALITY

2.1 INSPECTION

Witness points

Stages: Give sufficient notice so that inspection may be made at the following stages:

- Damp-proof courses, in position.
- Flashings, in position.
- Bottoms of cavities, after cleaning out.
- Bottoms of core holes, before grouting.
- Control joints, ready for insertion of joint filler.

Hold points >

2.2 SAMPLES

Masonry unit samples

General: Submit face units of each type illustrating the range of variation available, including colour, texture, surface irregularities, defective arrises, and shape.

Number of each type: 6.

Facework sample panels

General: Provide in a suitable position a sample panel of each type of facework including face or pointing mortar.

Facework type: >

Location: >

Minimum size (face of panel) (mm): >

Footings: >

3 MATERIALS AND COMPONENTS

3.1 MATERIALS AND COMPONENTS

Steel components

Durability classification to AS 3700:

- Steel products (including reinforcement), at least: >

Connectors and accessories

Durability classification to AS/NZS 2699.2: >

Mortar materials

Cement type to AS 3972: GP.

White cement: Iron salts content ≤ 1%.

Lime: To AS 1672.1.
Sand: Fine aggregate with a low clay content and free from efflorescing salts, selected for colour and grading.
Sand for facework:
- Colour: >
- Grading: >
- Source: >

Admixtures: Do not provide admixtures.
Coloured mortar:
- Colour: >
- Location: >

Premixed mortar: >

Mortar mix table

Mortar type to AS 3700	Mortar proportions (cement:lime:sand)	Location
M3	1:0:5 + water thickener	Concrete or calcium silicate masonry
M4	1:0:4 + water thickener	Grouted and reinforced masonry
M4	1:0 – 0.25:3	Underpinning, high strength masonry
M3	1:1:6	Other masonry

Masonry units schedule

Attribute	Type of unit			
	Clay masonry unit	**Concrete masonry units**	**AAC masonry units**	**Calcium silicate masonry units**
Source				
Location				
Manufacturing dimensions (mm)				
Category				
Dimensional category				
Unconfined compressive strength (MPa)				
Coefficient of expansion (mm/m)				
Salt attack resistance category				

4 EXECUTION

4.1 CONSTRUCTION GENERALLY

Cleaning

General: Clean masonry progressively as the work proceeds. Clean facework to remove mortar smears, stains and discolouration. Do not use acid.

Concealed work

Joints: Cut flush, and leave unstruck.

Sills and thresholds

Bedding: Solidly bed masonry sills and thresholds and lay them so that the top surfaces drain away from the building. Set out so that no unit is cut smaller than 3/4 full width.

Sill units: >

Threshold units: >

Joints and cutting

Set out: Set out masonry with joints of uniform width and minimise cutting of masonry units.

Holes, sleeves and chases: Build in during erection.

Depth of raking (other than facework): >

Rods

76 mm high units: 7 courses to 600 mm.

90 mm high units: 6 courses to 600 mm.

190 mm high units: 3 courses to 600 mm.

Bonds

Single leaf: Stretcher bond.

Facework: Stretcher bond.

Existing work: >

Tolerances (mm): >

Built in steel door frames

Fill the backs of jambs and heads solid with mortar as the work proceeds.

5 COMPONENTS

5.1 FACEWORK

Single face walls

Location: >

Double face walls

General: Select face units for uniform width and double-face qualities in single leaf masonry with facework both sides. Before starting, obtain a ruling as to which is the preferred wall face, and favour that face should a compromise be unavoidable.

Location: >

Commencement

Position: Commence at least 1 full course for blockwork, or 2 full courses for brickwork, below adjacent finished ground level.

Perpends

Alignment: Vertically align perpends in alternate courses.

Perforations

Exposed: Provide solid face units where perforations would otherwise be visible.

Joints

Surface: Work with a jointing tool to a dense smooth surface, except where the surface is to be bagged.

Joint profile: >

Depth of raking: >

Colour mixing

General: Where the colour of the face units is visible, evenly distribute the colour range of units. Prevent colour concentrations and "banding".

5.2 SUBFLOOR

Bearer support

Piers: Support bearers on engaged and free standing masonry piers at 1800 mm maximum centres.

Engaged piers:

- Brickwork: 230 x 110 mm, bonded to walls.
- Blockwork: 390 x 90 mm, bonded to walls.

Height (mm)	Brickwork (mm)	Blockwork (mm)
< 1500	230 x 230	390 x 190
1500 – 2700	350 x 350	390 x 390

Free standing piers table

Access openings

General: In internal walls, leave door width openings beneath doorways to give access to underfloor areas.

Air vents type

Brickwork:

- Terra cotta: Perforated, 230 x 160 mm.
- Concrete framed: Bronze wire mesh in concrete frames, 470 x 160 mm.
- Cut brick: 2 cut bricks laid vertically and evenly spaced in a 230 mm wide x 2 course high opening, backed with bronze wire mesh built in.

Blockwork:

- Vent blocks: Purpose-made vent blocks.
- Concrete framed: Bronze wire mesh in concrete frame 390 x 190 mm.

Air vents location

Ventilation rate: At least 7300 mm^2 free ventilation area per linear metre of wall.

Location: Below damp-proof course, within 600 mm of corners, elsewhere as required by ventilation rate, to internal and external walls.

Cavity walls: Provide matching vents in the internal leaves located as near as practicable to the vents in the external leaves.

5.3 PRECAST DOOR THRESHOLDS

Location >

Profile >

Finish to exposed faces

Off steel forms.

Concrete mix

1:2:4 cement:coarse aggregate 10 mm size:fine aggregate.

5.4 DAMP-PROOF COURSES

Material

Standard: To AS/NZS 2904.

Type: >

Location

General: Provide damp-proof courses in the following locations, if applicable:

- Walls adjoining infill floor slabs on membranes: In the course above the underside of the slab in internal walls and inner leaves of cavity walls. Project 40 mm and dress down over the membrane turned up against the wall.
- Cavity walls built off slabs on ground: In the bottom course of the outer leaf, continuous horizontally across the cavity and up the inner face bedded in mortar, turned 30 mm into the inner leaf 1 course above. Project 10 mm beyond the external

slab edge and turn down at 45°.
- Masonry veneer construction: In the bottom course of the outer leaf, continuous horizontally across the cavity. Fasten to the inner frame 75 mm above floor level. Project 10 mm beyond the external slab edge and turn down at 45°.
- Internal walls built off slabs on ground: In the first course above floor level.
- At timber floors: In the first course below the level of the underside of ground floor timbers in internal walls and inner leaves of cavity walls.

Installation

General: Lay in long lengths. Lap full width at angles and intersections and at least 150 mm at joints. Step as necessary, but not exceeding 2 courses per step. Sandwich damp-proof courses between mortar.
Junctions: Preserve continuity of damp-proofing at junctions of damp-proof courses and waterproof membranes.
Location: At least 150 mm above adjacent finished ground level.
Lap sealing: >

5.5 CAVITY WALLS

Minimum cavity width

Masonry walls: 50 ± 10 mm.
Masonry veneer walls: 25 mm, between the masonry leaf and the loadbearing frame and 40 mm between the masonry leaf and sheet bracing.

Openings

Closure: Do not close the cavity at the jambs of external openings.

Cavity fill

Height: Fill the cavity to 1 course above adjacent finished ground level with mortar weathered towards the outer leaf.

Flashings material

Standard: To AS/NZS 2904.
Type: >

Flashings location

General: Provide flashings and weatherings in the following locations, if applicable:
- Floors: Full width of outer leaf immediately above slab or shelf angle, continuous across cavity and up the inner face bedded in mortar, turned 30 mm into the inner leaf 2 courses above. Where the slab supports the outer skin and is not rebated, bed the flashing in a suitable sealant.
- Under sills: 30 mm into the outer leaf bed joint 1 course below the sill, extending up across the cavity and under the sill.

- Over lintels to openings in cavity walls: Full width of outer leaf immediately above the lintel, continuous across cavity, turned 30 mm into the inner leaf 2 courses above. Extend at least 50 mm beyond the lintels.
- Over lintels to openings in masonry veneer construction: Full width of outer leaf immediately above the lintel, continuous across cavity. Turn up against the inner frame and fasten to it. Extend at least 50 mm beyond the lintels.
- At abutments with structural frames or supports: Vertical flashing in the cavity using 150 mm wide material, wedged and grouted into a groove in the frame opposite the cavity.
- At stiles where cavities are closed: Full height flashing extending 75 mm beyond the closure into the cavity, interleaved with the sill and head flashing at each end. Fix to frame stiles.

Flashings installation

General: Sandwich flashings between mortar except where on lintels or shelf angles.

Lap sealing: >

Pointing: Point up joints around flashings, filling voids.

Weepholes

Form: Open perpends.

Maximum spacing: 720 mm.

Location: Provide weepholes to external leaves of cavity walls in the course immediately above flashings, and cavity fill, and at the bottoms of unfilled cavities.

Wall ties

Durability classification to AS/NZS 2699.1: >

Classification to AS/NZS 2699.1	Service conditions
Light duty	Masonry veneer
Medium duty	Normal cavity construction and at abutments
Heavy duty	Cavities > 60 mm wide

Wall ties category table

Wall ties installation

Fixing of masonry veneer ties at abutments:
- To timber frames: Clouts or integral spikes.
- To concrete: Masonry anchors.
- To steel frames: >
- To structural supports: >
- Special requirements for cyclone areas: >

Spacing: >

Embedment of wall ties

Cavities > 60 mm wide: 75 mm minimum.

Flexible wall ties

Type: Where ties or anchors extend across control joints, use ties or anchors which do not impair the effectiveness of the joint.

5.6 AUTOCLAVED AERATED CONCRETE WALLS

General

Type: A proprietary system which

- has a current Australian Building Product and Systems Certification Scheme certificate; or
- has a current appraisal report issued by the CSIRO Building Products and Systems Appraisals stating that the system is suitable for use in walls in buildings.

5.7 CONTROL JOINTS

Filler material

Type: Provide compatible sealant and bond breaking backing materials which are non-staining to masonry. Do not provide bituminous materials with absorbent masonry units.
Foamed materials: Closed-cell or impregnated, not water-absorbing.
Bond breaking materials: Non-adhesive to sealant, or faced with a non-adhering material.
Primer: Required.
Slip joints: >

Installation

Cleaning: Clean joints thoroughly before sealing.
Joint width:

- Vertical joints: >
- Horizontal joints: >

Sealant depth: 0.67 - 1.0 times joint width.

Control joints schedule

Joint code	Joint width	Joint filler		
		Primer	Backing rod	Sealant

6 REINFORCEMENT

6.1 BED JOINT REINFORCEMENT

Reinforcement

Material: Galvanized welded wire mesh.
Width: Equal to the width of the masonry leaf, less 15 mm cover from each exposed surface of the mortar joint.

Installation

General: Lap 450 mm at splices. Fold and bend at corners so that the longitudinal wires are continuous. Stop 200 mm short of control joints.
In brickwork: Extend 450 mm beyond each side of openings.
Location:

- In third bed joint above bottom of wall.
- In second bed joint below top of wall.
- In first 2 bed joints above and below openings.
- In first 2 bed joints above and below head and sill flashings to openings.

Maximum vertical intervals: 500 mm.

6.2 REINFORCED MASONRY

General

Designation: Masonry strengthened with embedded steel reinforcement, other than bed joint reinforcement.
Reinforcement tolerances: >

Cleaning core holes

Blockwork: Provide purpose-made cleanout blocks or machine cut a cleaning hole at the base of each reinforced core. Locate on the side of the wall which is to be rendered or otherwise concealed. Cover the hole with formwork and grout the core.

Bond beams

Type: Provide bond beams made from purpose-made hollow concrete blocks with reinforcement grouted in place.
Reinforcement: Two 12 mm diameter galvanized rods.

7 LINTELS

7.1 STEEL LINTELS

Material

Type: Mild steel galvanized to AS/NZS 4680, minimum coating mass 600 g/m^2. Do not cut after galvanizing.

Durability classification to AS/NZS 2699.3: >

Steel lintels schedule

Lintel dimensions (mm)	Maximum span (mm)
75 x 8	
75 x 10	
90 x 90 x 6	
90 x 90 x 8	
100 x 100 x 6	
100 x 100 x 8	
150 x 90 x 8	
150 x 100 x 10	

Cold-formed lintels

Type: Proprietary flat-base type designed to AS/NZS 4600.
Type tests: Required.

Tension bars

Type tests: Required.

Installation

General: Provide 1 lintel to each wall leaf. Do not cut on site. Keep lintels 6 mm clear of heads of frames. Pack mortar between the angle upstand and supported masonry units.
Minimum bearing each end:
- Span ≤ 1800 mm: 150 mm.
- Span > 1800 mm, ≤ 3000 mm: 230 mm.

Propping: To prevent deflection or excessive rotation, temporarily prop proprietary cold-formed lintels until the masonry reaches its required strength.
Minimum propping period: 3 days.

8 FINISHES

8.1 BAGGING

Dry bagging

Preparation: Cut joints flush before bagging.
Application: Apply laying mortar to the surface using a hessian bag or similar. Flush up irregularities, but leave the minimum amount of mortar on the surface.

Bibliography

AACA (Architects Accreditation Council of Australia) (1993), *NCSA 01: National competency standards in architecture – The competency standards*, AACA, Canberra.

ABCB (Australian Building Codes Board) (1999), *The BCA – Future directions*, ABCB website, www.abcb.gov.au.

– (2000), *BCA96* (to Amendment 6), CCH Australia, Sydney; Amendment 7 print CanPrint Communications, Canberra; electronic, Standards Australia Electronic Services, Sydney.

ABS (Australian Bureau of Statistics) (1992-93), 8676.0 *Technical services, Australia*, ABS, Canberra.

– (1995-96), 8693.0 *Consultant engineering services*, ABS, Canberra.

ACCC (Australian Competition and Consumer Commission) (1997), *$1,000,000 plus damages for victims of "cynical" home design and "predator" builder*, ACCC media release, 16 January, Sydney.

ACE (Association of Consulting Engineers) et al (1987), *CPI: Project specification – A code of procedure for building works*, BPIC (Building Project Information Committee), London.

ACEA (Association of Consulting Engineers Australia) (1993), *Practice note D.4 – Liability for use of out-of-date or superseded standards (revision 1)*, ACEA, North Sydney.

Achenbach, Joel (1999), *'Is the information age making us any wiser?'*, International Herald Tribune, 16 March.

Ackerman, James S (1986), *The architecture of Michelangelo*, Penguin Books, Harmondsworth.

ACSE (Association of Consulting Structural Engineers of NSW) (1991), *Concrete specification*, ACSE, Sydney.

Adams, Stewart (1989), *Practical buildability*, CIRIA (Construction Industry Research and Information Association) Building Design Report, CIRIA/ Butterworths, London.

AIA (American Institute of Architects) (1997), AIA document A201-1997: *General conditions of the contract for construction*, AIA, Washington.

– (1999), *CAD layer guidelines*, AIA, Washington.

AIQS (Australian Institute of Quantity Surveyors) (1990), *Australian standard method of measurement*, edition 5, AIQS/MBA, Canberra.

AJ Bicknell & Co. (1979 reprint), *Bicknell's Victorian buildings – Floor plans and elevations for 45 houses and other structures*, Dover Publications, NewYork.

Alberti, Leone Battista (*see* Rykwert, Joseph).

Alexander, Christopher (1964), *Notes on the synthesis of form*, Harvard University Press.

Allen Allen & Hemsley (1994), *New works and services contract*, Allen Allen & Hemsley, Sydney.

ALLC (Australian Language and Literacy Council) (1996), *Putting it plainly*, AGPS, Canberra.

anon (1974), *The new English bible*, Harmondsworth, Penguin Books.

anon (1994), 'Breaking the chain: materials producers', *Building*, 2 September.

anon (1995), 'Foresite', *Building*, 15 September.

anon (1996), 'Materials bar coding urged', *Building*, 22 November.

anon (1998), 'The pay-if-paid clause that may not pay', *The Construction Specifier*, June.

anon (1999b), 'The gradual greening of MASTERSPEC', *Environmental Building News*, September.

anon (2000), 'Safe design update', *NOHSC News*, Vol 15, No. 2.

ASTM (American Society for Testing and Materials) (1996), ASTM E 1557 *standard classification for building elements and related sitework – UNIFORMAT II*, ASTM, West Conshohocken, PA.

– (1999), C140-99b *standard test methods of sampling and testing concrete masonry units and related units*, ASTM, West Conshohocken, PA.

Barbour Report (1994), Barbour Index, London.

Barrymore, Karina (1995), 'Cautious reactions as both sides digest landmark decision', *Australian Financial Review*, 24 March.

Bartholomew, Alfred (1846), *Specifications for practical architecture*, John Williams & Co, London.

Bennett, David (1980), *Brooking on building contracts*, Butterworths, Melbourne.

Bennett, John et al (1996), *The University of Reading design and build forum report: designing and building a world-class industry*, Centre for Strategic Studies in Construction, University of Reading, Reading.

BIA (Building Industry Authority) (1998), *Approved document B2: durability*, BIA, Auckland.

Billingham, Erica (1994), 'Breaking the chain: contractors', *Building*, 16 September.

Bingham, Tony (1995), 'Liability between the lines', *Building*, 26 May.

– (1996), 'When it's wise to supervise', *Building*, 9 August.

Bishir, Catherine et al (1990), *Architects and builders in North Carolina: a history of the practice of building*, The University of North Carolina Press, Chapel Hill.

Boersma, Joh (1970), *Athenian building policy from 561/0 to 405/4 BC*, Wolters-Noordhoff Publishing, Groningen.

Borkowski, Andrew (1994), *Textbook on Roman law*, Blackstone Press, London.

Braunfels, Wolfgang (1972), *Monasteries of Western Europe – The architecture of the orders*, Thames and Hudson, London.

Brennan, Patrick (1993), 'Substitutions', *The Construction Specifier*, October.

Briggs, Martin (1927), *The architect in history*, Oxford University Press, Oxford.

Brown, Andrew (1994), 'Performance specified work', *the architects' journal*, 9 March.

BSC (Building Services Corporation) (1994), SAA MP70-1993 *The plain English home building contract*, BSC/Standards Australia, Sydney.

BSI (British Standards Institute), London:

– BS 1192 (various) *Construction drawing practice.*

– BS 5750:1:1987 (withdrawn) *Quality systems – Specification for design/development, production, installation and servicing.*

– BS 7750:1992 (withdrawn) *Environmental management systems.*

– BS 8000 (various) *Workmanship on building sites.*

Buick, J and Jevtic, Z (1995), *Cyberspace for beginners*, Icon Books, Cambridge.

Bullen, David (1991), *An architect's guide to building codes and standards*, The American Institute of Architects, Washington.

Burd, Edward (1994), 'Quality without control', *the architects' journal*, 9 March.

Burford, Alison (1969), *The Greek temple builders at Epidauros*, Liverpool University Press, Liverpool.

Burke, Noel (2000), Personal communication, August.

Bushnell, Christopher (1999), 'Today's tools for "writing" specifications', *The Construction Specifier*, August.

Bycroft, Peter (1988), *Some shortfalls in briefing*, BSFA (Building Science Forum of Australia) Workshop.

– (1989), *Behind the facade: what post-occupancy studies tell us about the quality of contemporary architecture*, RAIA (Royal Australian Institute of Architects) National Convention, Canberra.

CABO (Council of American Building Officials) (1995), *Model energy code*, CABO, Falls Church, VA.

Castello-Cortes, Ian (managing editor) (1994), *The Collins factopedia world reference atlas*, HarperCollins Publishers, Pymble.

Centre for Design at RMIT University, and Society for Responsible Design (1999), *EcoSpecifier handbook*, RMIT, Melbourne.

CIC-1 (*see* RAIA Practice Services, 1998e, AN15.11.100).

CIDA (Construction Industry Development Agency) (1994), *Building best practice in the construction industry – Overview for clients, consultants and contractors*, Standards Australia, Sydney.

CIS (Construction Information Systems), Milsons Point:

– *NATSPEC BASIC* (updated every January and July).

– *NATSPEC BUILDING* (updated every January and July).

– *NATSPEC SITE+STRUCTURE* (updated every January and July).

– *NATSPEC SERVICES* (updated every January and July).

– *NATSPEC DOMESTIC* (updated every January and July).

– *NATSPEC DOMESTIC – The better homes specification* (updated annually).

– *NATSPEC Reference – The national building and services reference specification* (updated annually).

Clarke, Somers and Engelbach, R (1990 reprint), *Ancient Egyptian construction and architecture*, Dover Publications, New York.

Colvin, HM (ed) (1963), *The history of the king's works volume 1: the Middle Ages*, HMSO, London.

Committee of Inquiry into Australia's Standards and Conformance Infrastructure (1995), *Linking industry globally*, AGPS, Canberra.

Conant, Kenneth John 1979, *Carolingian and Romanesque architecture 800-1200*, Penguin Books, Harmondsworth.

Cook, Andy (1997), 'Top marks', *Building*, 7 February.

Coomber, Matthew (1994), 'Breaking the chain: subcontractors', *Building*, 23 September.

Cooper, Susan and Page, Magdalen (1989), *Instructions for consumer products*, HMSO, London.

Cordell (1992), *Cordell's building cost guide – Commercial industrial – NSW*, Cordell Building Publications, Sydney.

Cornick, Tim (1991), *Quality management for building design*, Butterworth Architecture, Guildford.

Coulton, JJ (1977), *Greek architects at work – Problems of structure and design*, Paul Elek, London.

Cox, Stanley and Hamilton, Alaine (eds) (1991), *Architect's handbook of practice management*, RIBA Publications, London.

– (1995), *Architect's job book*, RIBA Publications, London.

Crawford, Marshall, Cann, John and O'Leary, Ruth (1997), *Uniclass: unified classification for the construction industry*, RIBA Publications, London.

Cruickshank, Dan and Wyld, Peter (1975), *London: the art of Georgian building*, The Architectural Press, London.

CSI (Construction Specifications Institute), Alexandria, VA:

– (1995), *MasterFormat.*

– (1996a), *Manual of practice – Construction documents fundamentals and formats module.*

– (1996b), *Manual of practice – Construction specifications practice module.*

– (1996c), *Manual of practice – FF/120 methods of specifying.*

– (1997), *SectionFormat.*

– (1998), *UniFormat.*

– (1999), *PageFormat.*

– (various), *Uniform drawing system.*

C21 (*see* DPWS, 1999a).

CUPDR (Committee on Uniformity of Plumbing and Drainage Regulations in NSW) (1992), *New South Wales code of practice – Plumbing and drainage*, CUPDR, Sydney.

Davie, Tom (1994), 'Out of date out of mind', *SPECnews*, October.

– (1995a), 'Bills of quantities' (letter), *SPECnews*, October.

– (1995b), 'Builders' liability', *SPECnews*, July.

– (1995c), 'Can you afford bills of quantities?', *SPECnews*, April.

Department of Defence (1992), *User guide*, DoD, Canberra.

Department of Fair Trading NSW (1998a), *Home building contract*, DFT, Sydney.

– (1998b), *Minor works building contract*, DFT, Sydney.

– (1998c), *Renovations contract*, DFT, Sydney.

– (1998d), *Swimming pool contract*, DFT, Sydney.

Dilke, OAW (1987), *Reading the past: mathematics and measurement*, British Museum Press, London.

DIST (Department of Industry Science and Technology) (1995), *Linking industry globally – Report of the committee of inquiry into Australia's standards and conformance infrastructure*, AGPS, Canberra.

DOE (Department of the Environment) and The Welsh Office (1992), *The building regulations 1991: approved document C: site preparation and resistance to moisture*, HMSO, London.

Doggart, John (1996), ECD Energy & Environment, Personal communication.

Dorsett, Charles (1993), 'Substitution and/or equal', *Specifics*, CSI, San Francisco Chapter newsletter Vol XXIX No 10.

Dorter, John and Sharkey, John (1981), *Building and construction contracts in Austtralia – Law and practice*, The Law Book Company, Sydney.

– (1995), *Building and construction contracts in Australia*, 2nd edition, The Law Book Company, Sydney.

Downes, Kerry (1987), *Sir Christopher Wren: the design of St. Paul's Cathedral*, Trefoil Publications, London.

DPWS (Department of Public Works and Services NSW) (1993), *Concise method of measurement of construction works*, DPWS, Sydney.

– (1996), *C21 construction contract conditions*, DPWS, Sydney.

– (1997), Documentation taskforce, Personal communication, Sydney.

– (1999a), *C21 construction contract*, DPWS, Sydney.

– (1999b), *PWM 0632 minor works contracts government agency version*, DPWS, Sydney.

Dundin, Joseph (1988), 'A hundred years (or so) of AIA standard documents', *Architecture*, October.

Eagleson, Robert (1990), *Writing in plain English*, AGPS, Canberra.

Eggleston, AS (1955), *The practising architect*, Melbourne University Press, Melbourne.

Ellickson, Dale (1995), 'The legalities of CAD', *Architecture*, March.

English, Shirley (1999), 'Builder alleges £20 million overspend', *The Times*, 16 January.

Evans, Robin (1995), *The projective cast – Architecture and its three geometries*, The MIT Press, Cambridge, MA.

Ferguson, Ian (1989), *Buildability in practice*, Mitchell, London.

Ferguson, Ian and Mitchell, Eric (1986), *Quality on site*, BT Batsford, London.

Franklin, Kim (1999), 'When the duty to warn can become a liability', *the architects' journal*, 25 March.

Frisch, Teresa (1971), *Gothic art 1140-c. 1450: sources and documents*, Prentice-Hall, Englewood Cliffs, NJ.

Frommel, CL (1994), 'Reflections on the early architectural drawings', in Millon, Henry A and Lampugnani, Vittorio M (eds) (1994), *The Renaissance from Brunelleschi to Michelangelo – The representation of architecture*, Thames and Hudson, London.

Gagarin, Michael (1986), *Early Greek law*, University of California Press, Berkeley, CA.

Gelder, John (1992), 'Specifying quality management', *Specifier*, Vol 1, No 2.

– (1993a), 'Australian standards in specifications', *Specifier*, Vol 2, No 1.

– (1993b), 'Use new contracts with care', *Specification News*, October.

– (1993c), 'What, no StandardsMark?', *Specification News*, October.

– (1994a), 'American ways of specifying', *Specification News*, April.

– (1994b), 'Avoiding repetition in the documents', *Australian Building News*, July.

– (1994c), 'Beware of scope of (the) work(s)', *Specification News*, July.

– (1994d), 'Services culture shock', *SPECnews*, October.

– (1994e), 'The contract documents as a project management tool', in proceedings of *National Construction and Project Management Conference*, Sydney, February.

– (1994g), 'Tropical timber certification', *SPECnews*, October.

– (1995a), 'Life without bills', *SPECnews*, April.

– (1995b), 'The specification as a building control tool', in proceedings of *Pacific Rim Conference of Building Officials*, Darwin, May.

– (1995c), 'Two-part documents', *SPECnews*, October.

– (1996a), 'Building modelling: a historical perspective', *INCIT 96 Proceedings, International Construction Information Technology Conference*, Sydney, April.

– (1996b), 'Gelder's guide to everyday standards', *SPECnews*, January.

– (1996c), 'Performance specifications: some issues for specifiers', BSFA (Building Science Forum of Australia) *Seminar Papers: Performance Specifications*, BSFA, Sydney, June.

– (1997a), 'Building product standards and conformance in Australia', *ATLAS of building standards and product quality*, CIS (Construction Information Systems), Milsons Point.

– (1997b), 'Can we have some order, please?', *SPECnews*, February.

– (1997c), 'How to get your products specified. How to get the specification to stick', BRANZ breakfast, Sydney, July (summarised in 'A healthy and informative Sydney breakfast', *Build*, October).

– (1999), 'DES 30: specifying for waste minimisation', *BDP Environment Design Guide*, BDP (Australian Council of Building Design Professions), Melbourne, November.

Gimpel, Jean (1988), *The medieval machine – The industrial revolution of the middle ages*, Pimlico, London. *See also* Waugh, Teresa.

Golden, Jean and Furey, William (1994), 'Safety considerations in job specifications', *The Construction Specifier*, November.

GPC (Government Paint Committee) (1993), *Specification E-15/4 – 1993 Full gloss enamel (buildings): interior only (manufacturer's colour range)*, APAS, East Melbourne.

Granger, Frank – trans of Vitruvius (1998 reprint), *On architecture: books I-V*, Harvard University Press, Cambridge, MA.

Grant, Michael (1984), *The history of ancient Israel*, Phoenix, London.

Gray, Don (1995), 'A message from the Chairman', *TAS – The Australian Standard*, Standards Australia, July.

Gross, James (1990), 'International harmonization of standards', *Proceedings ASCE (American Society of Civil Engineers) Structures Conference*, May.

– (1993), 'Building codes and standards in international trade', *Proceedings CSI 37th Annual Convention*, June.

Gulvanessian, Haig (1999), BRE *IP 3/99: structural Eurocodes – Their adoption in the UK*, BRE (Building Research Establishment), Watford.

Hale, JR (1977), *Renaissance fortification – Art or engineering?*, Thames and Hudson, London.

Hall, Dennis J (1999), 'The US national CAD standard: its history and the implications', *The Construction Specifier*, August.

Hall, Francis (1994), 'Specifying for quality', *the architects' journal*, 8 June.

Hambly, Maya (1982), *Drawing instruments – Their history, purpose and use for architectural drawings*, RIBA Drawings Collection, London.

Handisyde, Cecil (1976), *Everyday details*, The Architectural Press, London.

Harrison, HW and Keeble, EJ (1983), *Performance specifications for whole buildings: report on BRE studies 1974-1982,* BRE (Building Research Establishment), Watford.

Hartog, Peter (1992), 'Approved equal fairytale', *Specifier*, Vol 1, No 2.

– (1994a), 'Facades need depth of understanding', *Building Owner and Manager*, August.

– (1994b), Personal communication, Bangkok.

Haselberger, Lothar (1985), 'The construction plans for the Temple of Apollo at Didyma', *Scientific American*, December.

– (1995), 'Deciphering a Roman blueprint', *Scientific American*, June.

Haslem, Benjamin (1996), 'The great construction debate', *The Weekend Australian*, April 20-21.

Hawkes, Nigel (1999), 'How Nasa put its foot in Mars blunder', *The Times*, 2 October.

Heineman, Tom (1994), 'The imperative spirit', *The Construction Specifier*, February.

– (1995a), 'Intent is not enough', *The Construction Specifier*, January.

– (1995b), 'Shades of meaning', *The Construction Specifier*, March.

– (1999a), 'The role of the specifications consultant: part 1', *The Construction Specifier*, February.

– (1999b), 'The role of the specifications consultant: part 2', *The Construction Specifier*, March.

Hill, Donald (1984), *A history of engineering in Classical and medieval times*, Croom Helm, Beckenham.

HM Treasury Procurement Group (1997), *Procurement guidance No 3: appointment of consultants and contractors,* HM Treasury, London.

Horn, Walter and Born, Ernest (1979), *The plan of St. Gall: a study of the architecture & economy of, & life in a paradigmatic Carolingian monastery (3 volumes),* University of California Press, Berkeley, CA.

Humphries, P (1983), *Castles of Edward the First in Wales*, HMSO, London.

Hunt, AS and Edgar, CC – trans (1977 reprint), *Select papyri II*, Harvard University Press & William Heinemann Ltd.

Hutchinson, Maxwell (1993), 'The need to stick to the specification', *the architects' journal*, October.

ICE (Institution of Civil Engineers) (1993), *The new engineering contract*, Thomas Telford, London.

IH/RIBA (Institute of Housing/Royal Institute of British Architects) (1983), *Housing design brief*, IH/RIBA, London.

IHT (1999), *International Herald Tribune*, 23 January.

Info-Link (2001), *The book 2001 – Building products & suppliers*, Info-Link/Reed Business Information, Sydney, infolink.com.au.

ISO (International Organization for Standardization), ISO 9699:1994 *Performance standards in building – Checklist for briefing – Contents of brief for building design*, Geneva.

Jackson, Larry (1993), *Plain English drafting*, CPLI Better Contracts Through Better Documents Seminar, Sydney.

JAS-ANZ (Joint Accreditation System of Australia and New Zealand) (1999), Personal communication, July.

JCC (Joint Contracts Committee) (1994a), *Building works contract – JCC-C 1994 with quantities*, JCC, Canberra.

– (1994b), *Building works contract – JCC-D 1994 without quantities*, JCC, Canberra.

JCT (Joint Contracts Tribunal) (1993), *Practice note PN 25 Performance specified work*, RIBA Publications, London.

– (1998a), *PWQ98 standard form of building contract*, RIBA Publications, London.

– (1998b), *WCD98 standard form of building contract with contractor's design*, RIBA Publications, London.

Jeppesen, Kristian (1958), *Paradeigmata – Three mid-fourth century main works of Hellenic architecture reconsidered*, Jutland Archaeological Society Publications Vol. IV, Aarhus University Press, Aarhus, Denmark.

Joyce, Raymond (1995), *The construction (design and management) regulations 1994 explained*, Thomas Telford, London.

Jupp, James (1995), 'Ethnic and cultural diversity in Australia', *Year Book Australia 1995*, ABS, Canberra.

Kalms, Bryan (1990), *Guide to specification writing*, AGPS, Canberra.

Kasimer, Joseph and Miller, Lesa (1992), 'Using proprietary specifications', *The Construction Specifier*, July.

Keegan, John (1993), *A history of warfare*, Pimlico, London.

Keenan, Chris (1999), *UK metrication history and status*, US Metric Association website.

Kent, Robin (1985), 'The tabernacle', *RIBA Journal*, May.

Kerr, James Semple (1986), *Fort Denison*, National Trust NSW, Sydney.

Klein, Manfred et al (1991), *Type & typographers*, Architecture Design and Technology Press, London.

Knoop, Douglas and Jones, GP (1967), *The mediaeval mason – An economic history of English stone building in the later Middle Ages and early modern times*, Manchester University Press.

Kostof, Spiro (ed) (1977), *The architect*, Oxford University Press, New York.

Lambert, Susan (1983), *Printmaking*, HMSO, London.

Lawrence, Arnold (1983), *Greek architecture*, Penguin Books, Harmondsworth.

MacDonald, William (1982), *The architecture of the Roman Empire – Volume I*, Yale University Press, New Haven, Connecticut.

Mango, Cyril (1972), *The art of the Byzantine Empire 312-1453: sources and documents*, Prentice-Hall, Inc, Englewood Cliffs, NJ.

– (1978), *Byzantine architecture*, Faber and Faber, London.

MBA/CIESG (Master Builders Australia/Construction Industry Engineering Services Group) (1994), *Sub-contract – SC JCC-C 1994 MBA/CIESG*, Canberra.

McCallum, Neil; West, Rob; Olson, Don and Novak, Robert (1998), 'Construction mock-ups: lessons learned', *The Construction Specifier*, February.

McCann, John (1995), *Clay and cob buildings*, Shire Publications, Princes Risborough, UK.

McClendon, Susan (1997), 'Performance specifying for design-build', *The Construction Specifier*, May.

McConnell, John (1989), *The seven tools of TQC*, Delaware Books, Dee Why.

McEvedy, Colin (1972), *The Penguin atlas of modern history (to 1815)*, Penguin Books, Harmondsworth.

McGregor, Colin (ed) (1997), IC*IS Report No 1: a comparison of national master specification systems*, ICIS, Ede, The Netherlands.

– (1999), Personal communication.

Merritt, Chris (1995), 'Builders beware: new rule on liability', *Australian Financial Review*, 24 March.

Miller, Barbara (1998), 'Product certification: the mark of quality, the assurance of performance', *The Construction Specifier*, September.

Millon, Henry A (1994), 'Models in Renaissance architecture', in Millon, Henry A and Lampugnani, Vittorio M (eds) (1994), *The Renaissance from Brunelleschi to Michelangelo – The representation of architecture*, Thames and Hudson, London.

Milton, Hans (1992), *SAA HB23-1992: thesaurus of Australian construction terms,* Standards Australia and NCRB (National Committee on Rationalised Building), Sydney.

– (1994), *SAA HB50-1994: glossary of building terms,* NCRB, Standards Australia and Suppliers Index, Sydney.

Mindt, Bettina (1995), *ICIS Report No 2: specification systems in European construction,* ICIS, Geneva.

Minogue, Ann and Klein, Rudi (1997), 'Designer trouble', *Building,* 3 October.

Morgan, Morris Hicky – trans of Vitruvius (1960 reprint), *The ten books on architecture,* Dover Publications, New York.

Morris, G (1996), 'Modelmaker', *Metropolis,* January/February.

Morris, Roderick (1999), 'Baroque: ridiculously sublime', *International Herald Tribune,* 16-17 October.

Mortlock, Bryce and Milton, Hans (1989), *NATSPEC classification,* NATSPEC, Milsons Point, Sydney.

Nall, David (1999), *SpecPack,* Conventionall, Windsor.

Nawar, George and Zourtos, Kyrie (1994), 'And the walls came tumbling', *SPECnews,* October.

Nelson, Charles (1988, out of print), *Checkit! 1 project quality record,* RAIA Practice Services, Melbourne.

– (1991), *Risk management for design professionals,* RAIA National Education Division, ACT.

– (1998), *ABC2 architects model quality management system* (version 3.1h), Building Technology Pty Ltd, Melbourne.

NIBS (National Institute of Building Sciences) (1993), *Construction criteria base on compact disc – The world of construction information* (Brochure), NIBS.

Nielsen, Mary and Nielsen, Kris (1981), *Risks and liabilities of specifications in reducing risk and liability through better specifications and inspections,* American Society of Civil Engineers, New York.

NPWC/NBCC (National Public Works Conference/National Building and Construction Council) Joint Working Party (1990), *No dispute: strategies for improvement in the Australian building and construction industry,* NPWC, Canberra, May.

NPWD (NSW Public Works Department) (1994), *Draft procedure for technical specifications,* NPWD, Sydney.

NSW Government (1993), *Local government (approvals) regulation 1993,* NSW Government Gazette No 73.

Olmstead, Patrick (1996), 'An open look at "closed" specs', *The Construction Specifier*, April.

Ostler, Timothy (1994), 'Does quality assurance work?', *the architects' journal*, 16 February.

Parand, Foroutan (1999), 'What's in IT for the construction industry', *Constructing the future*, Winter, Issue 2.

Partridge Partners (1996), *The building material ecological sustainability index* (Vol 1, issue 2), Partridge Partners, Sydney, May.

PCA (Property Council of Australia) (1998), *Project contract PC-1,* Construction Publications Pty Ltd, Avalon.

Peaslee, Horace (1939), 'Streamlined specifications', *Pencil Points*, August.

Peck, Michael (1995), 'Qualification based selection', *SPECnews*, April.

PHCA (Parliament House Construction Authority) (1990), *Project parliament: the management experience*, AGPS, Canberra.

Pilley, John (1996), 'Letter to the editor', *The Australian Standard*, October.

Plommer, WH (1973), *Vitruvius and later Roman building manuals*, Cambridge University Press, London.

PMI (Project Management Institute Standards Committee) (1996), *A guide to the project management body of knowledge*, PMI, Upper Darby, PA.

Pollitt, JJ (1965), *The art of ancient Greece, 140-31 BC: sources and documents*, Prentice-Hall Inc., Englewood Cliffs, NJ.

– (1983), *The art of Rome c. 753 BC-AD 337: sources and documents*, Cambridge University Press, Cambridge.

Poole, Michael (1995), 'Can you afford not to have a bill of quantities?', *SPECnews*, July (also, *see* Letters in the same issue).

Preston, Jerry (1999), 'Change orders', *The Construction Specifier*, January.

Professional Specifications (1998), Personal communication, Brisbane.

Pugh, Francis (1987), 'Design, engineering and the art of drawing', in Papadakis, Andreas (ed) (1987), *Architectural design profile 70: engineering & architecture*, Academy Group, London.

Qaisar, Ahsan Jan (1988), *Building construction in Mughal India*, Oxford University Press, Delhi.

QAS (Quality Assurance Services) (2000), *Product buyers guide*, August, Sydney.

Queensland Department of Public Works (1996), Personal communication, November.

RAIA/MBA (Royal Australian Institute of Architects and Master Builders Australia) (1997), *SBW-2 lump sum contract for simple building works*, RAIA/MBA, Canberra.

RAIA Practice Division, RAIA Practice Services, Melbourne:

– (1984), *Advisory note AN16.01.601, Don't do the work of the builder*, July.

– (1986, now obsolete), *Advisory note AN15.01.402, When a varied contract is a repudiated sub-contract.*

– (1989), *Advisory note AN10.01.608, Beware of the fine print*, October.

– (1990a), *Advisory note AN04.02.101, Quality control in the preparation of specifications*, March.

– (1990b), *Advisory note AN16.01.605, Builders supervise building work – architects do not*, March.

– (1992), *Advisory note AN10.01.610, Site safety*, August.

– (1993a), *Advisory note AN10.01.101, Scope of the architect's services*, October.

– (1993b), *Advisory note AN14.02.100, Warranties and guarantees*, July.

– (1993c), *Advisory note AN15.01.100, Supply of documents under standard lump sum contracts*, July.

– (1994a), *Advisory note AN11.02.100, Issuing CAD files to builders and consultants*, October.

– (1994b), *Advisory note AN14.03.100, Specification writing technique*, January.

– (1995a, now obsolete), *Advisory note AN02.03.701, Architectural services cost survey.*

– (1995b), *Advisory note AN16.01.100, Shop drawings*, July.

– (1995c), *Advisory note AN16.10.100, Nominated or identified sub-contracts*, January.

– (1996a), *Advisory note AN02.03.100, Fee guide no. 8*, July.

– (1996b), *Advisory note AN05.01.001, The use of Australian standards in architectural practice*, April.

– (1996c), *Advisory note AN05.01.100, Architects' library of standards*, April.

– (1996d), *Advisory note AN05.01.600, Australian standards – before you quote them, understand them*, April.

– (1996e), *Advisory note AN10.01.110, Australian standards and client and architect agreements*, April.

– (1996f), *Advisory note AN10.01.601, Do you 'include by reference'?*, April.

– (1996g), *Advisory note AN14.01.100, The use of Australian standards*, April.

– (1997a), *Advisory note AN08.02.606, Beware indemnity provisions that exceed your professional indemnity insurance cover*, October.

– (1997b), *Advisory note AN10.01.603, Extra liability and professional indemnity insurance*, October.

– (1997c), *Advisory note AN17.01.100, The building operation and routine maintenance manual*, July.

– (1998a), *Advisory note AN02.01.100, Retention of documents*, January.

– (1998b), *Advisory note AN10.03.100, The design brief*, July.

– (1998c), *Advisory note AN11.01.101, Project records*, January.

– (1998d), *Advisory note AN14.01.612, Occupational health and safety legislation*, July.

– (1998e), *Advisory note AN15.11.100, CIC-1 Construction industry contract with interleaved notes*, October.

– (1999), *Advisory note AN14.03.103, Specification preliminaries*, January.

– (2000), *Advisory note AN15.01.603, Incorporation of government policy*, October.

– (2000), *Advisory note AN15.02.102, Amendments to SBW- 2 agreement for alterations and additions and where the proprietor remains in occupation*, January.

– (2001), *Newsletter*, January.

Selector.com (2000), selector.com, Melbourne.

Read, REH (1986), *British statutes relating to fire (1425-1963)*, Building Research Establishment, Borehamwood.

RIBA (Royal Institute of British Architects) with ACE et al (1993), *CPI: production drawings: a case study (3 vols)*, BPIC, London.

RIBAIR (Royal Institute of British Architects Indemnity Research) (1989), *Quality assurance*, RIBAIR, London.

RICS (Royal Institute of Chartered Surveyors) (1998), *Standard method of measurement of building works*, 7th Edition, RICS, London.

Rogers, Louise (1993), 'Material changes', *RIBA Journal*, November.

Rosen, Harold (1999), *Construction specifications writing – Principles and procedures*, John Wiley & Sons, New York.

Rosen, Harold and Heineman, Tom (1990), *Construction specifications writing – Principles and procedures*, John Wiley & Sons, New York.

Rothenberg (1995), 'Ensuring the longevity of digital documents', *Scientific American*, January.

Rykwert, Joseph et al - trans of Alberti (1988), *Or the art of building in ten books*, The MIT Press, Cambridge, Mass.

SA (Standards Australia), Sydney:
- AS 1100 (various) *Technical drawing.*
- AS 1214-1983 *Hot-dip galvanized coatings on threaded fasteners (ISO metric coarse thread series).*
- AS 1225-1984 *Clay building bricks.*
- AS/NZS 1388: 1994 *Guidelines for technical information for building and construction products.*
- AS 1428 (various) *Design for access and mobility.*
- AS/NZS 1530.3-1999 *Methods for fire tests on building materials, components and structures – Simultaneous determination of ignitability, flame propagation, heat release and smoke release.*
- AS 1604-1997 *Timber – Preservative-treated – Sawn and round.*
- AS/NZS 1677.1:1998 *Refrigerating systems – Refrigerant classification.*
- AS 1742.2–1994 *Manual of uniform traffic control devices – Traffic control devices for general use.*
- AS/NZS 1768: 1991 *Lightning protection.*
- AS 1884-1985 *Floor coverings – Resilient sheet and tiles – Laying and maintenance practices.*
- AS 1897-1976 *Electroplated coatings on threaded components (metric coarse series).*
- AS 2050-1995 *Installation of roof tiles* .
- AS 2124-1992 *General conditions of contract.*
- AS/NZS 2312:1994 *Guide to the protection of iron and steel against exterior atmospheric corrosion.*
- AS 2688-1984 *Timber doors.*
- AS 2700-1996 *Colour standards for general purposes.*
- AS 2818-1993 *Guide to swimming pool safety.*
- AS 2896-1998 *Medical gas systems – Installation and testing of non-flammable medical gas pipeline systems.*
- AS 2924-1987 *Decorative thermosetting laminated sheet.*
- AS/NZS 2924.1:1998 *High pressure decorative laminates – Sheets made from thermosetting resins – Classification and specifications.*
- AS/NZS 2924.2: 1998 *High pressure decorative laminates – Sheets made from thermosetting resins – Determination of properties.*
- AS 2925 (various) *Methods of testing thermosetting laminated sheet.*

– AS/NZS 3000:2000 *Wiring rules.*

– AS/NZS 3017:1996 *Electrical installations – Testing guidelines.*

– AS/NZS 3018:1997 *Electrical installations – Domestic installations.*

– AS/NZS 3500 (various) *National plumbing and drainage code.*

– AS/NZS 3500.3.2: 1998 *National plumbing and drainage code – Stormwater drainage – Acceptable solutions.*

– AS/NZS 3500.5:2000 *National plumbing and drainage code – Domestic installations.*

– AS 3600-1994 *Concrete structures.*

– AS 3610-1995 *Formwork for concrete.*

– AS 3700-1988 *Masonry in buildings.*

– AS 3700-1998 *Masonry structures.*

– AS 3798-1996 *Guidelines on earthworks for commercial and residential developments.*

– AS 3828-1998 *Guidelines for the erection of building steelwork.*

– AS 3854.2-1990 *Performance standards in building – Principles for their preparation and factors to be considered.*

– AS/NZS 3905.2:1997 *Quality system guidelines – Guide to AS/NZS ISO 9001, AS/NZS ISO 9002, and AS/NZS ISO 9003 for construction.*

– AS 3958 (various) *Ceramic tiles.*

– AS 4000-1997 *General conditions of contract.*

– AS 4058-1992 *Precast concrete pipes (pressure and non-pressure).*

– AS 4072.1-1992 *Components for the protection of openings in fire-resistant separating elements – Service penetrations and control joints.*

– AS 4100-1998 *Steel structures.*

– AS 4120-1994 *Code of tendering.*

– AS 4145.2-1993 *Locksets – Mechanical locksets for doors in buildings.*

– AS/NZS 4284:1995 *Testing of building facades.*

– AS 4300-1995 *General conditions of contract for design and construct.*

–AS 4303-1995 *General conditions of subcontract for design and construct.*

– AS/NZS 4455:1997 *Masonry units and segmental pavers.*

– AS 4460-1997 *Australian standard for construction of premises processing meat for human consumption.*

– AS/NZS 4680:1999 *Hot-dipped galvanized (zinc) coatings on fabricated ferrous articles.*

– AS 4901-1998 *Subcontract conditions.*

– AS 5601-2000 *Gas installations.*

– AS/NZS ISO 8402:1994 *Quality management and quality assurance – Vocabulary.*

– AS/NZS ISO 9000 (various) *Quality management and quality assurance standards.*

– AS/NZS ISO 9001:2000 *Quality management systems – Requirements.*

– AS/NZS ISO 9002: 1994 *Quality systems – Model for quality assurance in production, installation and servicing.*

– AS/NZS ISO 14000 (set) *Environmental management standards set.*

– AS/NZS ISO 14001:1996 *Environmental management systems – Specification with guidance for use.*

– AS CA55-1970 *Code of recommended practice for the design and installation of bituminous fabric roofing.*

– SAA HB23-1992 (*see* Milton, Hans).

– SAA HB40-1992 *The Australian refrigeration and air conditioning code of good practice.*

– SAA HB50-1994 (*see* Milton, Hans).

– SAA HB57-1994 *User's manual for lump sum contract based on AS 2124-1992.*

– SAA HB70.2-1994 *HIA Quality management system – Project procedure and inspection manual.*

– SAA MP70-1993 – *see* BSC (Building Services Corporation).

SAA (2000), Personal communication, July.

Salzman, LF (1952), *Building in England down to 1540 – A documentary history*, Oxford.

SBW-2 (1997) (*see* RAIA/MBA).

Scientific Services Laboratory (1999), *Register of accredited products – Fire protection equipment*, Port Melbourne.

Scott, John (1964), *A dictionary of building*, Penguin Books, Harmondsworth.

– (1984), *A dictionary of building*, Penguin Books, Harmondsworth.

Senior, Derek (1964), *Your architect*, Hodder and Stoughton, London.

Shakespeare, William (1598), *'King Henry IV part 2'*, in Evans, G Blakemore (ed) (1974), The Riverside Shakespeare, Houghton Mifflin Co., Boston.

Shrive, Charles (1993), *Improving engineering consultants' specifications*, 37th Construction Specifications Institute (CSI) Convention, Houston, CSI, Alexandria, VA.

SPECnews (2000), 'Greenhouse initiatives', Winter.

SSL (Scientific Services Laboratory) (1999), *SSL register of accredited products – Fire protection equipment*, SSL, Melbourne.

STABU (Stichting Standaardbestek voor de Burger-en Utiliteitsbouw) (1993), *Standard building specifications*, STABU, Ede, The Netherlands.

Standen, David (1993), *Construction industry terminology*, RAIA Practice Services, Melbourne.

– (1995), *Construction industry specifications*, RAIA Practice Services, Melbourne.

Stevens, Garry (1990), *The reasoning architect – Mathematics and science in design*, McGraw-Hill, New York.

Stitt, Fred A (1994), *Production systems for architects and designers*, Van Nostrand Reinhold, New York.

Sydgood Pty Ltd (1991), *St George Hospital Kogarah – Building survey for planned preventative maintenance system*, Paddington.

Thompson, Jessica Cargill (1996), 'Latham team attacks professionals', *Building*, 12 January.

Thorne, Robert (1987), 'Paxton and prefabrication', in Papadakis, Andreas (ed) 1987, *Engineering & architecture*, Academy Group, London.

Thumtack, Tom (pseud.) 1914, 'Specifications', *Architecture and Building*, April (quoted in Peaslee – *see* Peaslee, Horace).

Tibbs, Judith (1998), 'New study compares design-build with other project delivery methods', *The Construction Specifier*, December.

Tilley, Paul and McFallan, Stephen (1999), *Design and documentation quality – Perceptions of contractors and designers*, CSIRO DBCE (Division of Building Construction and Engineering), Canberra.

Tooley, Angela (1995), *Egyptian models and scenes*, Shire Publications, Princes Risborough, UK.

Tyrril, John (1995), 'The new engineering contract, Institution of Civil Engineers, Thomas Telford 1994', *The Architecture Show*, March/April.

Uher, Thomas (1994), *Partnering*, AIPM (Australian Institute of Project Management) seminar, AIPM, Sydney.

Vitruvius (*see* Granger, Frank, and Morgan, Morris Hicky).

Vognild, Richard (1998), 'Coming soon: the international building code', *The Construction Specifier*, September.

Wallis, Bob (1995), 'Development of a national construction law for Vietnam', *Proceedings Pacific Rim Conference of Building Officials*, Darwin.

Warner, Denis and Peggy (1995), 'The way we lived', *The Australian Magazine*, 19-20 August.

Waugh, Teresa – trans of Gimpel (1983), *The cathedral builders*, Pimlico, London.

Westgate, Rona (1995), 'A reasonable question', *Building*, 15 September.

Wheildon, Colin (1990), *Communicating or just making pretty shapes*, Newspaper Advertising Bureau of Australia, North Sydney.

Whiteley, Adam (1998), 'Writing specifications', *the architects' journal*, 22 January.

Willis, Anne-Marie, and Tonkin, Cameron (1998), *Timber in context – A guide to sustainable use*, CIS, Milsons Point.

Willis, Christopher and Willis, J Andrew (1997), *Specification writing for architects and surveyors*, Blackwell Science, London.

Willis, Robert and Clark, John Willis (1988 reprint), *The architectural history of the University of Cambridge* (3 volumes), Cambridge University Press, Cambridge.

Wolfe, Sheldon (1994), 'Public works bidding', *The Construction Specifier*, April.

Worsley, Giles (1991), *Architectural drawings of the Regency period 1790-1837*, AIA Press, Washington.

Wright, Terry et al (1998a), *NATSPEC Classification 2 draft*, CIS website, www.cis.asn.au.

– (1998b), 'Classifying information', *SPECnews*, February.

Wyatt, David J (1998), 'ASTM: a century of standards', *The Construction Specifier*, June.

Zantkuijl, HJ (date not known), *Bouwen in Amsterdam* (publisher not known).

Index